The Missing Link

The Emergence of Man

The Missing Link

by Maitland A. Edey
and the Editors ·
of Time-Life Books

TIME
LIFE
BOOKS

Time-Life International
(Nederland) B.V.

The Author: MAITLAND A. EDEY is former Editor of TIME-LIFE BOOKS. He is the author of *The Northeast Coast* in The American Wilderness series, and as Editor of the LIFE Nature Library he produced the book *Early Man*. He has travelled widely in Kenya, Tanzania and Uganda.

The Consultants: SHERWOOD L. WASHBURN is Professor of Anthropology at the University of California at Berkeley, and BERNARD CAMPBELL is Professor of Anthropology at the University of California at Los Angeles.

The Cover: Six million years ago Australopithecus, man's closest ancestor, emerges tentatively from the forest to the grassy reaches of Africa's savannas, a major step towards manhood for this link between tree-living animals and modern humans.

The cover scene, as well as those on pages 8 and 21-31, has been re-created by superimposing paintings of Australopithecines on photographs of landscapes like those that existed when some Australopithecines lived.

Contents

Introduction

For 2,500 years, since the inscription "know thyself" was placed above the lintel of the Temple of Apollo at Delphi in ancient Greece, wise men have known that man's salvation lies in self-knowledge. Religion, philosophy and, more recently, psychology are man's attempts to explain both his own nature and the development of his personality. But however much we have learned from such deliberations, something has been missing. This is an understanding of man's inborn nature—the genetic endowment—that man brings from his most distant past. This endowment consists of the biological roots of human anatomy and behaviour: the framework within which the environment would shape his nature. Only an understanding of the time before recorded history could tell us about the kind of creature we once were, and how we came to be the way we are.

The search for the prehistory of the human species began, like so much else in science, as an aspect of man's insatiable curiosity. No practical results were foreseen by those who undertook with so much patience and devotion the exploration and excavation of distant places: only the prize of new knowledge, which in its own way is one of man's greatest treasures. And this knowledge has given us a fresh perspective of ourselves. Within the last 15 years extensive new fossil and archaeological evidence has thrown valuable light on the whole process of man's evolution. Today, the buried record can tell not only of ancient man's bones and material culture—his stone and bone tools—but of something more: his environment, his diet and, finally, even his social life and behaviour. It is knowledge of man's prehistoric behaviour that is perhaps most significant. Though only a fraction of human behaviour is "programmed" by genetic inheritance, these innate characteristics of the nervous system provide the framework within which that behaviour is realized, and give that framework its extraordinary potential. The educability and behavioural flexibility of human beings are products of their past. The growing knowledge of how we survived in the past will prove to be critical in our attempts to survive in the present.

This book reviews the evidence for the earliest phase of man's evolution: the crucial period when he separated from the common ancestor he shares with the African apes. It is crucial because it was during this period that the most significant characteristics of man's anatomy and behaviour were evolved. It may also have been a very precarious time, for our ancestors, vulnerable creatures as they were, might easily have suffered in competition with other animal species and become extinct. But we know they did not: our ancestors survived by a unique adaptation to their environment—they left the friendly forest of their predecessors and became biped vegetarians, scavengers and hunters, in open country. They were not just social, but cultural animals —something altogether new in the world of nature.

The peculiar quality of this ancestral species from man's most distant past—this missing link—is pieced together in this book from the evidence of excavation, and from observations of the behaviour of our living primate relatives. We see the peculiar adaptation called human taking form before our eyes, man himself in the making. The emergence of man can be interpreted as a process in which a unique animal became conscious of itself and its place in nature. This book is a further step in this evolutionary progress towards self-knowledge.

—Bernard Campbell

Chapter One: The Ancestor

Almost human but still ape-like, the borderline Australopithecus peers from the African forest as it looked two million years ago.

I would not be ashamed to have a monkey for my ancestor, but I would be ashamed to be connected with a man who used great gifts to obscure the truth.
—T. H. Huxley, defending Darwin's theory against the attack of Anglican Bishop Samuel Wilberforce

Despite its scientific subject and all the jawbreaking names that will be found in it, this book should be regarded as a detective story. It has a central problem that demands to be solved, and a number of clues that accumulate as the story develops. Like any good mystery, this one starts with a body: the human body —your body, mine, the body of the man down the street. The most arresting thing about the human body is that it is unique. There is nothing like it in the world—no other one that combines the attributes of thinking, talking, always walking on its hind legs, making things with its hands, enjoying binocular colour vision; no other that depends primarily on its cultural adaptations rather than on its physical adaptations to enable it to get along. The problem: how did man get that way? Where on earth did this odd body come from?

"Who am I?" Every thinking man asks himself that question at one time or another. It is the most profound and interesting question there is. The answer he gets will depend on whom he asks. I, for example, have a name that identifies me to strangers—if they should ask. But my name is less useful to my postman than the number of the street I live in. At my bank I am known by an account number, at my office by a National Insurance number.

These numbers say a little about me, but very little. They reveal that I do have a bank account and a job, but their real purpose is to locate me, to find me among millions of other faceless numbered human beings so that money can be given to me or taken away, and parcels delivered. But what about *me*? My passport gives a few clues. It reveals that I am a man, that I am six feet tall and have brown eyes.

Six feet. Tall, but not exceptionally so. About four inches taller than the average American male, several inches taller than the average human. How did I get that way, since my parents were both short, three of my grandparents very short? Was my fourth grandparent—my father's father—responsible? I don't know. He abandoned his wife a few years after he married her, became a drifter and died a suicide before my father was 20. My family preferred not to talk about him.

Three of my grandparents had blue eyes. Did that little-known suicide bequeath me my pair of brown ones? What else? I sometimes wonder. That sense of helplessness and futility that hits me when things are going particularly badly?

I do not know. All I can say is that I am a product of those people, a new combination of inherited bits and pieces contributed by all of them, shaped by the environment I live in—an environment that is partly of my own making, partly theirs, and partly their parents' and their parents' parents'. When I ask myself who I am, I am forced to look back past a mother and father, whom I knew intimately, through eight great-grandparents who are little more than names, to 16 great-great-grandparents whose names I do not even know. Back and back my thoughts spin. There was a time a thousand years ago when I had countless ancestors all contemporary with one another, spread over much of the world. If no relative had ever married another, no matter how distantly related, there would have been more than a thousand million of them, all alive at approximately the same time. But the entire population of the world was only about 280 million then; second, third and fourth cousins have always been intermarrying without realizing their kinship. Therefore, to compensate for that overlap, let me arbitrarily divide that huge ancestral horde, not by two or by ten but by a thousand. That still leaves a million individuals, all different, all my ancestors, walking the earth at the time of William the Conqueror.

My people are mostly of English, Scottish and Dutch extraction. But what does that mean? Nothing, really, because a thousand years ago many of them were still being called Saxons and Picts. Since at that time 90 per cent of all the people in western Europe were peasants, it is certain that nearly all my ancestors were peasants too, unable to read or write, fearsomely ignorant and superstitious, routinely and appallingly cruel, ravaged by hard work and disease, many of them toothless by 25, dead by 35. I feel little kinship with them, but I *am* them. I carry their genes in my body—my looks, my shape, my desires, my predisposition to certain ailments, my ways of thinking—something from every one of them.

Go back another thousand years. I almost certainly pick up a good many Romans and Greeks as forebears, in addition to Semitic Middle-Easterners, some farther-eastern Tartars, some Egyptians, some black Africans. But the great bulk of my ancestors were surely the ancestors of those medieval European peasants. Two thousand years ago they were tribesmen, living in forests and along rivers, primitive agriculturalists. Civilization may have advanced elsewhere but not much of it had rubbed off on them. By modern standards they were savages. Some of

them may never have had their hands on a piece of metal. But still my ancestors.

Go back again, a hundred thousand, five hundred thousand, a million years. I continue to encounter ancestors, but now they no longer look like people. Compared to mine, their brains are small, their minds dim, their thought processes hopelessly narrow. If they can talk at all, they do so in a most primitive way. If they wear anything, they wear skins. Some of them may not even know the use of fire and may have subsisted all their lives on berries, roots and small animals like frogs and lizards caught and eaten raw. My ancestors.

Who am I? I am all these people, for a proper answer to the question must be a genetic one, an evolutionary one that enables me to relate myself to the whole of mankind, indeed to all living creatures. If I can do that, perhaps I can truly find out who I am, how it is that I have a large brain that enables me to write books—or, for that matter, to build cities, to invent and drive motor-cars, to fly to the moon. More basically, perhaps I can learn how it is that I can walk upright, speak and speculate about myself as I am doing now.

There was a time when my ancestors could do none of these things. They were not men, but ape-like creatures living in trees. Somehow they became men. There is a link somewhere in my ancestral line, on the very edge of human-ness, that connects creatures that clearly were men to creatures that clearly were not. The purpose of this book (also the problem in the detective story) is to identify that link, examine it in the light of a great deal of new evidence that has been pouring in during the last few years, and see if it, its forebears and its relatives can provide a satisfactory explanation of the process by which an ancestral arboreal ape became a man—in short, how I came to be what I am.

This in-between creature is the missing link of the book's title. It deserves that name for two reasons. It is indeed a link between men and non-men—more accurately, a series of links in a connected chain. It has been appropriate in the past to call it "missing" because all but the skimpiest evidence about it has been missing until comparatively recently. A great deal of important evidence is still missing, leading to some hot current debate as to how several closely related creatures connect on a chain so long-lost that its links have almost defied assembly. Recent discoveries and analysis now begin to make it possible to lay out some of those links next to one another, preparatory to clamping them irrevocably together. Each new find either will forge the existing links in the chain more strongly in a particular sequence or will require that they be pried apart and rearranged into another sequence—possibly even with the introduction of a new link or two.

Palaeoanthropology is the science of putting that chain together. For those engaged in it, this is a time of extraordinary interest. The best of scientists, at this very moment, cannot agree on some important matters. Several blockbusters have hit the field in the last couple of decades, and others are surely on the way. The layman, who is not competent to evaluate the finicky disputes of experts, can only wait for the blockbusters to fall, the dust to settle, and watch wide-eyed as the debris is sorted out. The book deals with the current state of that sorting-out process and makes a tentative identification of a missing link.

It all started in 1859 with the publication of Charles Darwin's *On the Origin of Species by Means of Natural Selection.* It is impossible today to re-create the atmosphere of intellectual and moral shock that swept England when the implications of that epochal book were made known. It was not that evolution of plants or animals was so hard to swallow. After all, man himself had been responsible, through selective breeding, for the evolution of a number of domestic animals and a great variety of crops. Then there were those peculiar dinosaur bones that people had begun digging up; they had to be explained—as did the growing evidence that the earth, instead of being 6,000 years old, as the churches taught, was hundreds of thousands, perhaps hundreds of millions of years old. No, those things were not really the problem. What was so hard to swallow was the suggestion that man was descended from a bunch of repulsive, scratching, hairy apes and monkeys.

Those awful monkeys! As one Victorian lady said: "My dear, let us hope that it is not true, but if it is let us pray that it will not become generally known." But it did become known. Those monkeys and apes seized the centre of the stage because of an exhaustive study of primates made by T. H. Huxley, a friend of Darwin and an ardent propagandist for Darwin's theory. Huxley's conclusion was that of all animals on earth the African great apes, the chimpanzee and gorilla, were most closely related to man. From this it followed that if prehuman fossils were ever found, they would lead to even older types that would turn out to be ancestral to both apes and men —and they would probably be found in Africa.

For the evolutionists this was a painful time. For all the logic of Huxley's view, there was an embarrassing lack of fossils resembling men in Africa or anywhere else to support it. When Darwin's book was published there was only one suspected fragment of this nature known in the entire world. It was part of a man-like skull found three years previously in a limestone cave in the Neander *thal*, or valley, in Germany. Obviously of human or human-related origin, this skull was still very odd in appearance. At the time, it was easier to regard it as a deformed specimen of modern man than to accept the possibility that human ancestors actually looked like that. A number of prominent scientists swept "Neanderthal man" under the rug.

But there was another problem that would not go away. If not Neanderthal man, then who was making the stone axe-heads and other crude implements that were turning up with perplexing frequency in river beds and in caves throughout western Europe? Furthermore, advances in geology were beginning to make it possible to calculate the age of some of these implements with some precision. Reluctantly the scientific world began to realize that many of them were more than 20,000 years old, some of them more than 100,000 years old. But no "proof" of their workmanship nor good identification of their makers could be made unless tools and human fossils could be found together in the same layers of debris in the floor of a cave or in the same gravel bed.

Ultimately a number of other Neanderthal fossils were discovered, as well as a great many of a later type, now known as Cro-Magnon man, after the name of the place in France where the first such remains were found. It became clear that men had existed in Europe for at least 100,000 years. Palaeoanthropology was now a respectable science with some hard fossil evidence. What it lacked was any idea of whence that evidence sprang. Who were the ancestors of Neanderthal man? What did *they* look like?

It remained for a young Dutchman named Eugene Dubois, an Army surgeon in Java, to supply an answer. Digging in 1891 and 1892 in the edge of a Javanese river bed rich in the fossils of extinct animals, he found part of a man-like skull and some teeth that were much more primitive than anything previously discovered.

Dubois gave his Java find the name of *Pithecanthropus erectus*, or erect ape-man, using two Greek words: *pithecos* (ape) and *anthropos* (man). Its age: about three-quarters of a million years.

This sudden septupling of the span of admitted human existence, particularly on the evidence of a single specimen from a remote part of the world, had the predictable result: few people believed it. But Dubois' find was getting confirmation elsewhere. A fossil of a somewhat similar type was uncovered in Germany; it was called Heidelberg man. Then, in the 1920s and 1930s, extensive digging in some hill caves near Peking turned up a large number of human fragments. These, though not as old or as primitive as Dubois' man, resembled him. Meanwhile, back in Java, others that were just as old and as primitive were found not far from where Dubois had worked. Their discoverer: G. H. R. von Koenigswald. Since then other finds and other living sites have been explored in Spain, France, Hungary, North Africa and East Africa. It gradually became clear that men—not apes or even ape-men, but *men*—were widely distributed half a million years ago throughout the warm and temperate regions of the Old World, in what is now considered to be a single species but with consid-

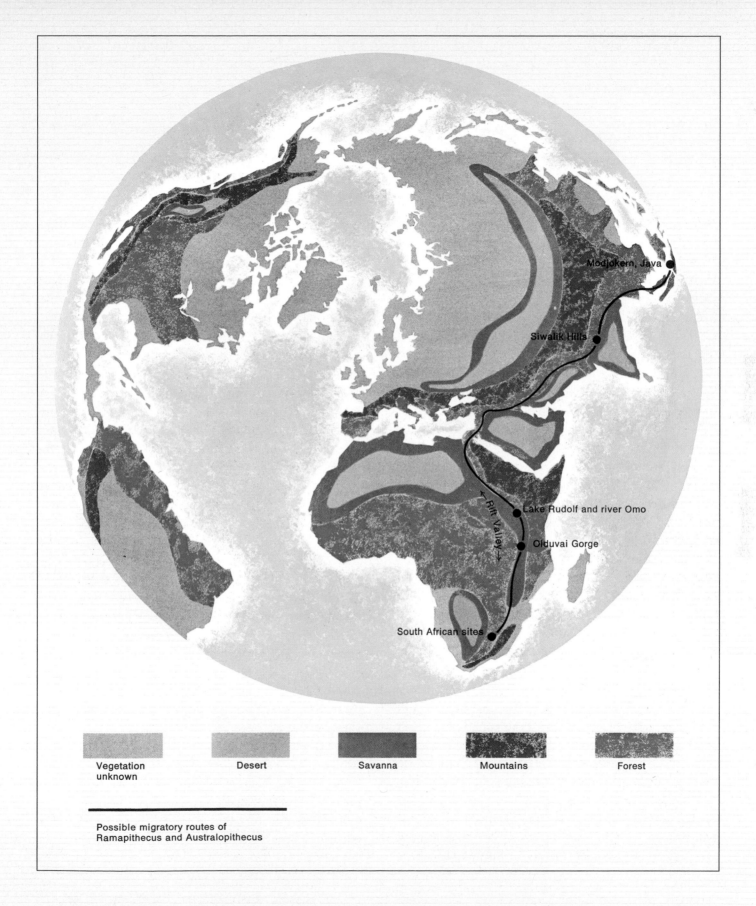

Modjokern, Java

Siwalik Hills

Lake Rudolf and river Omo

Rift Valley

Olduvai Gorge

South African sites

Vegetation
unknown

Desert

Savanna

Mountains

Forest

Possible migratory routes of
Ramapithecus and Australopithecus

erable local variation. In recognition of their unmistakably human characteristics, all of them were given a single scientific name: *Homo erectus*, which places them in the same genus as *Homo sapiens*, the name given to modern man.

This extension of the human line did not take place as neatly or as quickly as is told here. And, once again, it did not dispose of the central problem; it merely pushed it a few steps further back in time. There still remained the question: where did Homo erectus come from?

Again, an answer came from an unexpected and faraway place, this time South Africa. Raymond Dart, a Johannesburg anatomy professor with an interest in the past, had formed the habit of encouraging his students to send him rock fragments that appeared to contain fossils. In 1924, hearing from one student about a possible baboon skull from a limestone quarry at a place named Taung, he arranged to have several crates of rock from the quarry sent to him. When he opened the second crate, one of the first things that hit his eye was not a skull but the next best thing to it: an odd-shaped rounded piece of rock that appeared to be the mould of the inside of a skull. Later on, deeper in the box, he found the piece that this mould fitted into—the skull itself.

One look at this antique fragment sent Dart's mind rocketing off on an almost unimaginable tangent. Here was no fossil baboon. Its brain capacity seemed larger than a baboon's, and its face did not have the long projecting jaw and large canine teeth that clearly identify both existing and fossil baboons. This fossil had the smaller jaw and the more nearly vertical face plane of an ape. But apes live in tropical forests. There are no such forests in South Africa, nor

have there been for more than a hundred million years. Furthermore, after Dart had had time to study his find more carefully, he realized that there were things about its teeth that were more man-like than ape-like. Also, the foramen magnum—the hole in the base of the skull where the nerves of the spinal cord entered it—was so situated as to indicate that the creature stood erect. A man? No, it could not possibly be a man, it was too primitive, too small brained. A pre-man, then, a link with the ape past? Taking a deep breath, Dart announced to the world that he had found a hominid—a human ancestor that was not yet human. He named it *Australopithecus africanus* (southern African ape).

Once again, from the scientific world, resounding scepticism. This recurring suspicion may seem strange. After all, anthropologists spend their lives looking for increasingly primitive, ever more ape-like fossils. Why are they so reluctant to recognize one when it turns up? There are numerous reasons. For one, there are many false alarms. If this book were to catalogue all the mistaken claims about hominid fossils made by layman and expert alike, it would have to be far longer than it is. Also, there have been deliberate frauds. The most famous of these was Piltdown man, a modern human skull stained to look very old, then apparently planted by a prankish young anthropologist in a dig in England along with an ape's jaw whose teeth had been filed down to resemble those of a human. These were then exhumed by an amateur scientist, Charles Dawson, and for decades they gummed up the progress of anthropological thought to an extent that is hard to believe. The Piltdown find suggested that early men already had large brains but still had ape-like faces—a concept that satisfies modern human vanity, with its emphasis on the special quality of the human intellect. Dart's fossil was not so agreeable to human vanity. It suggested that just the opposite was true: that face and teeth began to become recognizably human while the brain was still very small.

Australopithecus fell onto this piece of scientific flypaper and got stuck there. A friend of Dart's, Robert Broom, tried to unstick it. In the redoubtable British scientific publication *Nature* he announced that Dart was right, and went off on his own to see if he could find other fossils of the same type. In due course he did. Others did also; at five major sites in South Africa they eventually turned up literally hundreds of Australopithecine fragments. There were enough of them, in fact, to permit speculation that there might be two species: a heavy-jawed "robust" type with extremely large molar teeth; and a smaller, slenderer "gracile" type with smaller molars. For a long time neither impressed scientists elsewhere. This was partly because Dart, the original discoverer, was unknown to the palaeoanthropological establishment, all of whose brightest stars were in England, France and Germany, partly because the brains of these fossil creatures were just not big enough to satisfy them. Perhaps Australopithecus was simply an aberrant chimpanzee.

The finds posed another problem as well. Instead of lying in strata that might give some clues to their age, the fossils were being blasted out of a concrete-like conglomerate of rocks and sand. There was no way of telling exactly how old they were. Even the animal fossils mixed in with them were no help; they were of extinct types. Broom had nothing to compare them to. Making a bold but shrewd guess, he announced that Australopithecus was probably two million years old.

More hoots of derision. Could anybody believe that an erect human ancestor with a brain scarcely bigger than a chimpanzee's had been running about in South Africa two million years ago? Apparently not. World War II came and went, and still Australopithecus was scarcely known to the world.

It was not until 1959 that this situation changed. The change came about through the efforts of Louis and Mary Leakey, a husband-and-wife team of East African anthropologists who until then had been engaged in one of the most persistent and unrewarding efforts in the history of anthropology. Leakey had a museum job in Nairobi, Kenya, but much of his spare time was spent in a dry river gulch in northern Tanzania, several hundred miles away. This place bore the name of Olduvai Gorge and was known as a rich source of animal fossils. But what particularly drew the Leakeys to it was the presence of extremely primitive stone tools scattered in the gorge, some lying loose on the ground, others working their way out of the valley's slowly eroding vertical walls.

Over a period of 28 years the Leakeys worked off and on at Olduvai. They had almost no money; the place was stiflingly hot. At first it took them days to get back and forth from Nairobi on the one brutally rough road to the gorge. Slowly, with the help of other experts, they came to know Olduvai's geological history, cut as it was by a now-vanished river to reveal strata of sediments and volcanic material lying one over another like layers in a birthday cake. From these various levels, over the years, they recovered an enormous number of animal fossils, identifying and classifying several hundred species, some of

The Human Family: Ape to Man in 14 Million Years

AUSTRALOPITHECUS BOISEI
The largest Australopithecine, Boisei was a vegetarian whose fossils have been found in East Africa. How well it could walk erect is uncertain. Not a human ancestor, it has been extinct about a million years.

RAMAPITHECUS
Known only from a few teeth and jaw fragments that date from about 9 to 14 million years ago, this ape-like creature is believed to be in the human line. No clues exist as to whether or not it walked erect.

AUSTRALOPITHECUS AFRICANUS/ HABILIS
Shown above is the form of Australopithecus found in both East and South Africa and thought to be man's immediate ancestor. Walking erect, using tools, its brain grew through time. Specimens less than two million years old are called Habilis: some experts consider them to be true men.

AUSTRALOPITHECUS ROBUSTUS
This creature, found only in South Africa, is closely related to Africanus. It actually was only slightly "robust" compared to Boisei. It, too, was a dead end and became extinct about one million years ago.

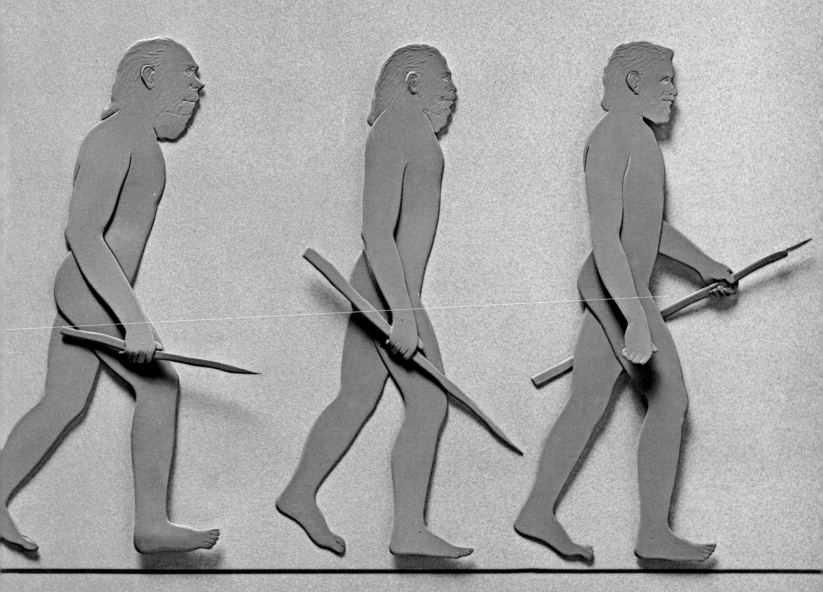

HOMO ERECTUS

The first true men began to emerge from the Africanus/ Habilis line between one and one and a half million years ago. Highly successful, Erectus made shelters and good tools, spread over Asia, Europe and Africa, and evolved into the modern species, Homo sapiens, about 200,000 years ago.

HOMO SAPIENS NEANDERTHALENSIS

Neanderthal man, who was of the same species as modern man, appeared some 100,000 years ago. He flourished as a big-game hunter during the last ice age, greatly expanding the area of human occupation, and made an extensive array of stone tools and weapons, and died out about 40,000 years ago.

HOMO SAPIENS SAPIENS

Modern man starts with Cro-Magnon man, who evolved out of Neanderthal populations scattered across most of the Old World. His oldest known remains go back at least 40,000 years. By 20,000 years ago he was producing cave paintings and other cultural objects in southern France and Spain.

them extinct, some of them hitherto unknown to science. But, aside from two small skull fragments and two teeth that they believed to be hominid, there was not a trace of man. Only those mocking stone tools. Who made them?

One afternoon Leakey was lying in his tent with a fever. His wife, unwilling to waste any of the precious time they had remaining at Olduvai, went for a last look-about before having to return to Nairobi. In the late-afternoon sun she spotted something sticking out of Bed I, the lowest layer of the gorge. It was part of a hominid face, with large brown teeth!

Quite rapidly thereafter several things fell into place. The Leakeys' "man" was found to be similar in some ways to the larger and more robust of the Australopithecine types that were turning up in South Africa. In fact, it was so much more robust that Leakey decided it was a different species. Other anthropologists agreed, and it was given its own name: *Australopithecus boisei*, after the Boise Fund, a financial source of support to the Leakeys. By a miracle of geology, Boisei could be dated. The skull lay just above a layer of volcanic ash, and the age of the ash could be determined. Boisei, it turned out, was about 1.75 million years old.

Naturally the South Africans were pleased by this news; it improved the age claims of their fossils. And it at last drew the attention of the palaeoanthropological world to Africa, which is what the South Africans had been trying to do for 30 years. It has been there ever since.

Does Australopithecus qualify as the missing link? In the sense that it appears to lie in the shadow land between man and ape, yes, it does. But Australopithecus comes in more than one model, and the names that scientists give those models have changed in the past and may change again—reflecting the confusing ways that one model dissolves gradually into another over a period of time.

So, for the moment, it will be prudent to defer the pinning of a final missing-link label on anything. Instead, now that we have Australopithecus roughly located in space (Africa) and in time (about two million years ago), let us take a look at this creature, bearing in mind that the following description is highly tentative. Some conclusions about its habits and appearance are better than others; none are undisputed.

He was not an ape. He lived either on the edge of the forest or right out in the open plain—but always within a day's walk of water. He was there a long time, certainly for two million years, maybe as many as five or six million years. He spent his days in small troops of males and females with their attendant children, plus some infants carried by their mothers. He walked on two legs like a man, but possibly with a less efficient gait than would be employed by a free-striding modern man. He could run well—and did, to catch lizards, hares, rats and other small prey. His world thronged with animals, as does East Africa today: enormous herds of antelope, zebra and other herbivores. He moved among them confidently, watching the smaller gazelles for a sick or crippled individual that he might kill, alert to the possibility of picking off a newborn calf. Deinotherium (a kind of extinct elephant), rhinoceros and the powerful black buffalo he discreetly sidestepped; they ignored him.

His world also contained the lion, leopard, hyena and a large, now-extinct sabre-toothed cat. All preyed on him from time to time. But he moved in compact bands, carried sticks, bone clubs and crudely chipped rocks, and exhibited a strong sense of group defence in the presence of any threat. All these things made it possible for him to go about his daily business of searching for roots, berries, insects and whatever larger game he could manage to catch, without serious threat from the big cats, which generally preferred to prey on antelope, as they still do. He may even have been able to drive single leopards or lions away from kills they had made; he could certainly have done this to hyenas, providing the hyenas were not too numerous.

He was agile, keen-sighted, alert. He was more intelligent than the baboons with which he shared the savanna, and had nothing to fear from them, although a single Australopithecine male would probably have been no match for a single male baboon. Australopithecus was lighter, less powerful, and lacked the baboon's murderous jaw and canine teeth.

The "gracile" male was between four and a half and five feet tall and weighed between 80 and 100 pounds; females were somewhat smaller. The colour of his skin is unknown, but it was probably lightly covered with fine hair. His face—let's face it—was very much that of an ape. His jaw stuck out more than a modern man's, but he had no chin to speak of. His nose was wide and flat, scarcely projecting. His forehead was low and sloping, the bony ridges over his eyes prominent. The top and back of his head were very small by modern standards—markedly so, compared to that large forward-thrusting face.

Judging by the smallness of his brain and by its presumed proportions, he could not talk, but he was certainly capable of a number of expressive sounds whose meanings others of his kind understood. He also communicated with them by means of a subtle and varied range of gestures, body movements and facial expressions.

Females, in common with many primates, had a monthly menstrual cycle. Unlike others, they were becoming sexually receptive throughout the month instead of during a few days at the peak of the cycle. There is no way of knowing whether couples paired up for a few weeks, for a year, for life; whether the most powerful male in the group kept several females to himself; or whether all members of the group were mutually and casually promiscuous. What is most probable is a slowly growing tendency for the fairly permanent pairing of couples since this is the basis of human family formation, is a fundamental aspect of human society and has obviously been developing over an immense span of time. Already, probably, there were the shadowy beginnings of a division of labour, which would also characterize later human social development. Males did more chasing around after game and the defending of the group. Females did more looking for roots and berries, and took care of infants. Food—whether caught or collected—was probably shared.

Where Australopithecines went at night is not known. In savanna areas with forest edges, they may well have taken to the trees to avoid predators, since night is the time when the latter would be most active, and the time the sleeping Australopithecines were most vulnerable. In open, drier country, where sizeable trees are lacking, it is possible only to speculate about how they survived. They may have made small thorn shelters for themselves, slept in rock crannies or caves whose entrances they may have blocked—or they may have simply slept in the open, relying on the relative scarcity of predators in arid country where game is not abundant, and on their own reputation as noisy and bothersome fighters that wise predators would do best to avoid.

Whatever they did, they survived, because I am here to talk about them. My ancestors.

This preceding thumb-nail sketch of Australopithecus is concise and clear enough. But is it true? That is a very large question that cannot be answered satisfactorily by simply lining out a confident description. There must be some proof in the form of arguments to back up that description, particularly since parts of it are given with more assurance than others. Awkward questions arise. For example:

1. There is a great deal of talk about the ages of these early men. How do we know that the estimates given are correct and not hundreds of thousands or millions of years out?

2. There is talk of "monkey-like" or "ape-like" or "man-like" characteristics. What are these? How are we going to be able to distinguish between a monkey, an ape and a man, particularly if all we have to go by is a handful of teeth?

3. The scientists say that we are descended from apes. What is the proof of that? Why not monkeys? And, by the way, just what does the monkey-ape-man family tree look like?

4. Last, and most important, how, when and why did apes and men begin to be different? If our ancestors were once four-legged creatures living in trees, what made them into two-legged creatures living on the ground?

Any book on the evolution of man is going to have to answer, in one way or another, questions like these. This book will attempt to deal with them by examining two kinds of evidence. One is fossils. The other is behaviour, the examination of living creatures, both men and their nearest primate relatives, for clues as to how we got the way we are today.

The Daily Life of Australopithecus in a Hunter's Eden

Euphorbias, trees that Australopithecus would have found familiar two million years ago, dot the shore of Uganda's Lake Edward.

Even now, some two million years after the age of the tool-using man-ape Australopithecus africanus, who is believed to have been ancestral to man, it is possible to reconstruct how he spent his days. He survived by collecting plant foods and occasionally hunting animals, and his fossil remains show that he lived on lake shores, along rivers or in the forest edges of open savanna. There are environments in East Africa today that closely resemble those places, and the combination painting-photographs on this and the following pages re-create not only the appearance of Africanus and his home ground but also the kinds of activities he is believed to have engaged in.

The scene above shows a typical lake-edge environment in Uganda. Much of the surrounding country is open savanna and is full of animals today—everything from small rodents to large herds of antelope and elephants. This combination of savanna and water edge is ideal for a hunting-gathering hominid, and is probably a near duplicate of the conditions that once existed in such now-dried-out places as Olduvai Gorge, Omo and the Lake Rudolf area, where Australopithecus is known to have lived.

An Existence
Tied to Water and Game

While one Australopithecus (far left) bends for a drink, two others make stone choppers. The nearby zebras and gazelles ignore them.

An Australopithecus pokes a twig into a termite mound to get insects to eat. Chimpanzees still do the same.

Early hominids lived out their lives cheek by jowl with the animals they hunted. Those same animals live with lions and other predators today, and show no fear of them, so long as they can keep them in view and do not feel they are being stalked. They can afford this sense of security because their speed and stamina enable them to escape most attacks if they see them coming. It is likely that prey animals similarly ignored the slow-running hominids, who probably had to corner or surprise them, or to single out a young or weak animal in order to make a kill.

Small water courses like the one shown here were good places for Australopithecines to gather. Here they might ambush game animals as they came to drink, and also find water-rounded stones from which flakes could be knocked to make crude but sharp stone tools. Areas rich in volcanic rock, quartz and chert would have been especially useful.

Clubs for Weapons, Branches for a Rain Dance

Shaking sticks, two scavenging hominids bully a timid cheetah, hoping to make it leave a kill it has just made.

Although abundant evidence survives that Australopithecines made and used a variety of stone tools, there is no trace whatsoever of the enormous number of perishable wooden implements they must have used. Chimpanzees brandish sticks and branches today. They have even been observed throwing these sticks and branches at baboons who try to take their food or get too close to a newborn chimp. Thus it seems logical to assume that the much more intelligent Australopithecines—capable of shaping stone choppers—would also have fashioned wooden clubs, spears, skewers and other sharp-pointed tools.

Weapons were certainly vital to them. The Australopithecines lived on the ground and had to compete there with many other dangerous carnivores. And it may be that they also used wooden implements for a quite different purpose—to brandish in a joyful celebration of rain, as chimpanzees are known to do today.

The great natural spectacle of a thunderstorm may have stimulated Australopithecines to run around waving branches in a rain dance.

26

A small band of foraging Australopithecines moves slowly across a stretch of Africa's fertile grassland, looking mostly for seeds and tubers, bu

also on the alert for grubs, hares, tortoises, fledgling birds or ostrich eggs. Much of their food was probably obtained in this hide and seek fashion.

The Relaxed Day
of a Primitive Predator

Drowsing in the heat of equatorial noontime, a male Australopithecus relaxes against a tree, while two females fuss over an infant.

Learning as they play, four spry youngsters chase one another around an acacia tree on the African savanna.

A fact not generally realized is that extremely primitive people have a great deal of leisure time. So it must have been with the Australopithecines. Their options were limited, and their needs few and easily met in a warm, benign environment. When there is enough to eat throughout the year, there is little else to do but sit around.

Such an abundance of free time led, of course, to socializing and to the development of complex inter-relationships between members of the groups. As intelligence grew, relationships increased in subtlety and intricacy. Meanwhile the periods of infancy, childhood and adolescence became longer and longer because of the need of an individual to learn more and more if it was to fit into a society of growing complexity. These tendencies, inherited from our ape-like ancestors, played a large part in the evolution of Australopithecines into humans.

At Nightfall, a Safe Resting Place in the Boughs

One of the mysteries of Australopithecine existence is how and where those creatures slept. Early in the history of their exploration of a life on the ground, they may have kept close to the forest edge—to the strips of woodland that lined lake and river —and retired to the trees for the night.

The trees were a refuge, for big cats and hyenas must have preyed on Australopithecus at night, just as they prey on modern savanna dwellers now. The Australopithecines would presumably have climbed up any type of tree that had branches strong enough to support them. A tree with a good broad branch and some surrounding lesser branches shooting off it could have provided a comfortable niche as is; but Australopithecus might also have gathered some grasses and leaves and fabricated a temporary nest for himself. His closest relatives, chimp and gorilla, both do —and there is no evidence he had the buttock calluses found on many of the monkeys that sleep wedged into the crotches of trees.

Later, after some millions of years' experience as ground dwellers—when they had become considerably larger brained and were beginning to be feebly cultured — Australopithecines undoubtedly started to make crude shelters of thorn hedge. Such hedges would have protected against nocturnal predators—and provided more comfort than trees.

With the light fading from the sky a group of Australopithecines leaves the dangerous

hadows creeping over an African plain and clambers into the trees for safety. Neither a hunting leopard nor a lion was likely to climb trees at night.

Chapter Two: The Evidence of Stones and Bones

A superb skull of a female Australopithecus africanus, found by Robert Broom in South Africa, lacks only teeth and lower jaw.

How often have I said it to you that when you have eliminated the impossible, whatever remains, however improbable, must be the truth.—Sherlock Holmes

It goes without saying that the study of prehistoric man is, of necessity, the study of his fossil remains. To begin to understand who our ancestors were and what they were like, we must be able to interpret the bits and pieces of them that are coming to the surface in increasing numbers. To do that, we must know how old they are. Strange shapes and sizes may suggest all sorts of intriguing hypotheses about who descended from whom. But those hypotheses—the relationship of one odd piece of bone to another—can be pinned down tight only by reliable dating.

In all matters of dating, what is most important is that everything fits together. The gradual revealing of the story of human evolution can be compared to the cleaning off of an old tapestry that has been covered with mud and dust. Modern man is at the top of the tapestry and his most primitive ancestors are woven into it near the bottom. The whole tapestry is fragile, increasingly so the nearer one gets to the bottom. It must be cleaned with great care to avoid destroying it. The cleaning process takes a long time. There is no assurance, if one should select a particular spot on the tapestry to begin picking away bits of dirt that that spot will turn out to have a relevant picture on it. It may be bare. It may have a hole in it. It may reveal only a fragment of one of the figures in the design, a fragment so small and cryptic as to be meaningless. It may reveal an entire figure, but a totally unexpected one whose presence cannot be explained until further parts are cleaned. But the position of every bit of design on the entire tapestry

—exactly where it is located in time with respect to all the other bits—is enormously important. Only in respect to all of the other elements can relevant deductions be made about specific fossils, and a proper human genealogy worked out.

The problem of age—of dating—is handled in three ways. The first is through geology, the study of the earth itself. This concerns itself with the location, size and nature of the various layers of clay, silt, sand, lava, limestone and other kinds of rock that constitute the earth's surface, and their relationship to one another. It notes that certain processes—such as erosion and the accumulation of layers of silt at the bottom of the sea and their compaction into rock again by heat and pressure—are taking place now at measurable rates, and assumes that those same processes took place at comparable rates in the past. Analysis of these layers—a scientific discipline known as stratigraphy—permits the working out of a rough picture of past earth history. From this the fossils found in different rock structures can be arranged in order of age.

The second way to determine age is through study of fossils themselves. They are not the same in different layers. They evolve through time and thus provide clues of their own, particularly if the time sequence can be worked out. The evolution of the horse, for example, is very well known through its fossils. Over a period of about 60 million years it developed from a four-toed animal the size of a cat to the large one-toed animal it is today. The numerous intermediate horse-fossil stages, located in various geological strata, tell this story with great clarity; any other animal or plant fossil that occurs in the same layer as one of those ancestral horses can be considered to be the same age. Once dated, a fossil can be used to help date another fossil, and so on.

By constant cross-checking and fitting together of enormous amounts of both rock and fossil evidence, science has been able to construct a rather detailed chronology of the past. But in this chronology specific dates are lacking.

The third technique supplies these dates. It is based on the knowledge that certain radioactive elements discharge energy at a constant rate, and in the process turn into something else. This discharge is known as the decay rate. Radium, for example, turns slowly but steadily into lead. Once this steady decay rate is known, it is only a matter of laboratory technique to determine how old a piece of radium is by measuring how much of it is still radium and how much is lead. One long-lasting radioactive substance is potassium 40. It is particularly useful because it is found in volcanic ash and lava. Fossils located in such volcanic rock or sandwiched between two layers of it can be dated with remarkable precision.

Given reliable clocks for measuring time, we can now turn with more confidence to primate fossils for an answer to the all-important question: how do we tell monkeys, apes and men apart? For present-day species this is no problem; all have evolved sufficiently so that they no longer resemble one another. But, since they all have a common ancestor, the farther back we go in time, the more their fossils begin to look alike. There finally comes a point when they are indistinguishable. It is that characteristic that makes it necessary to attempt the construction of a primate fossil family tree if we are ever going to find out what the hominid line of descent is.

In our attempt to sort out monkey, ape and hom-

inid fossils, most of our attention will be drawn to differences in jaws and teeth because they fossilize the best, and are often the only evidence we have. Teeth are by far the hardest and most enduring parts of the body. They are still recognizable as teeth, although they may be stained and worn, after literally millions of years of lying in the earth. Bone is more fragile. Sometimes it is well preserved and can stand the patient picking away by scientists of the surrounding bits of rock that cling to it. That first Taung skull found by Raymond Dart was durable enough to be worked over carefully by him for months before he got the face cleared. If was four years before he succeeded in getting the jaws apart so that he could examine the surfaces of the teeth.

Other fossil bones are not that durable. They collapse into powder as the surrounding material is removed. It is sometimes necessary to inject them with liquid plastic binder to hold them together before they can be worked loose from the ground.

Whether stony or powdery, fossils look different from rocks. Experts spot them instantly, and also have a phenomenal ability to tell them apart. Any professional, for example, can tell a monkey molar from an ape molar or a human molar. The cusps, or small bumps, on the grinding surface of a monkey molar usually number four, and are arranged in pairs. An ape or a fossil human usually has five cusps on some of its molars. More important, these are not neatly paired, but lie in a characteristic Y-shaped pattern. This Y-5 pattern is a primitive condition and is found in the common ancestor of both ape and monkey. Therefore, a 15-million-year-old jaw fragment with a couple of four-cusped molars still attached to it must be a monkey jaw and not the jaw of an ape or a hominid, both of which have retained the more primitive Y-5 cusp pattern.

Here is a useful bit of evidence for starting a sketch of the primate family tree, but it is negative evidence. It does not tell us where the monkey-ape fork should be pencilled in. It says only that the fork occurred before 15 million years ago. But how long before? Alas, one fossil can say only so much. For more evidence on the proper positioning of the fork, we will have to find older four-cusped monkey-type teeth, and then still older ones, and we may find ourselves erasing the pencil mark several times before we are through. Also, there is a built-in problem that evolution itself causes. The following imaginary exchange may help illustrate the difficulty:

"Those differences in cusps that you describe may be all very well for a fossil that is 15 million years old, but what about further back? Won't you eventually come to a time when monkey teeth begin to look more and more like the ancestral Y-5 teeth?"

"Yes."

". . . when you have a little something there that might or might not be a fifth cusp?"

"Yes."

"Well, what is it, an ape or a monkey?"

"If you go back that far, you will probably find other things about the size, the shape or the number of teeth that might help you, because cusp patterns are not the only dental characteristics that have evolved in the primate line. If you are lucky enough to find other bones, your identification job will be easier yet; the shoulder of a monkey, for example, is quite different from that of an ape. The latter can swing its arm more freely than a monkey can."

"All right—but if you keep going back, won't you

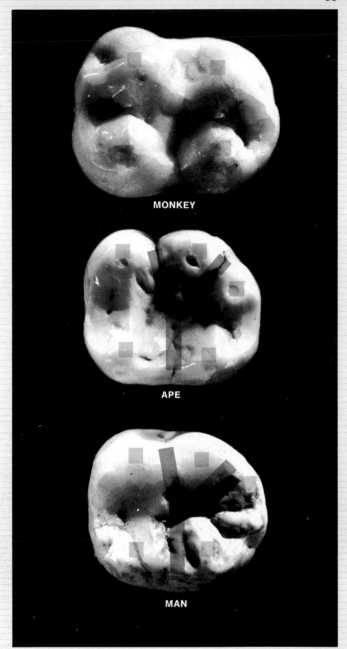

MONKEY

APE

MAN

The Y-5 tooth pattern—five cusps (green squares) separated by a Y-shaped valley that is indistinct in many teeth—distinguishes the molars of hominids and apes from those of monkeys. Monkey's molars never have more than four cusps and do not have the Y-shaped crease between them.

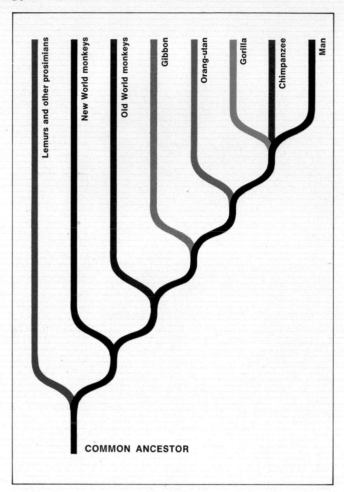

COMMON ANCESTOR

Man and all his primate relatives descend from a common ancestor, a small rat-like mammal. First the lemur and other prosimians split off, then the monkeys themselves split, some evolving in the New World, some in the Old. From the latter group, apes sprang. The first to separate from the early ape stock was the gibbon: the next was the orang-utan. A third ape-like line produced gorilla, chimpanzee and man.

get to an animal in which none of these characteristics are really clear?"

"Yes."

"Well. . .?"

"You're asking me to pinpoint the day when a monkey mother gave birth to an ape infant. There never was such a day. There was a very long period during which populations of ancestors with confusing blends of both monkey and ape characteristics existed. During this period they began to show local variations from one another in response to slight differences in their environment or their eating preferences. Even if we had a complete series of fossils, a couple of million of them representing millions of years of evolution, we could never pinpoint the very first one that could be called an ape or a monkey."

"Then, if you're making family trees, you won't be able to locate the branches very precisely, will you?"

"No, we won't. A so-called 'branch' actually covers a long period of time and its location will always be approximate. The best it can do is indicate about when, in a general population, certain differences began to appear, reflecting a difference in the environment or the feeding habits of the animals involved. Both geographical and behavioural factors tend to separate segments of a population. As a result they interbreed less and less. If a population should become truly divided—separated physically by a mountain range, an invasion of the sea or the slow spread of a desert area—they will not interbreed at all. Thus the process of becoming different will be accelerated. Ultimately two different kinds of animals will emerge where only one existed before. The emergence of four-cusped teeth in one group of primates is the result of some such process—along with a great many other important characteristics that reflect the variety of life styles that different kinds of monkeys and apes exhibit today."

That is a highly simplified account of how the process of speciation works. It is this process that has been responsible for the development not only of monkeys, apes and men but of all living things. It is also what makes possible the theoretical sorting out of a line of fossil teeth, their distinctive cusp pattern gradually becoming fainter and fainter until—when it disappears—a fairly good estimate can be made of about when apes and monkeys did begin to separate. Someday, if we find enough of those teeth, this will be done. If not, there are other jaw, or skull or skeletal characteristics—often combinations of them—that might also do the job. At any rate, whenever a convincing bundle of that kind of evidence can be assembled, one branch of a family tree can be pencilled in with some confidence.

Pencilling in the monkey-ape fork is just one of the important steps in tracing the descent of man. It is tempting to consider the relationship of men and apes as being even more important, since they are more closely related and their split more recent. Actually, every step along the way is important—none less so than any other—for the developments in each stage of the chain of life that leads to man have been made possible only by the developments that precede it. To understand precisely what men are, we must consider the entire chain. Specifically, we must go back more than 75 million years and take a careful look at certain rat-shaped, rat-sized, insect-eating mammals that were scuttling about on the ground in tropical rain forests.

Some of those early mammals began climbing up into the trees, presumably because of intense competition on the ground and because there was a rich source of food up there. Mammals like this called tree shrews survive today in certain places, relatively unchanged. Others did change. They changed so radically that their evolution is hard to believe. From those small insectivores sprang a whole array of prosimians, monkeys, apes and hominids—which raises another awkward question: If all those other creatures did succeed in evolving, why is that remnant of relatively unevolved tree shrew still hanging on in the modern world?

The reason is that they have not had to change much. Their environment has remained fairly static and they still fit it pretty well. It should be understood that there is no guarantee that a species will evolve. In fact, there are enormous pressures against evolution. Nature is conservative, and a population that is getting along well in a particular environment will tend to stay the way it is. Through the forces of natural selection the vast majority of its members will always resemble one another very closely. Nearly all of them approach a kind of "best available model" for that particular moment in time and for that particular place and that particular way of making a living. If they do not closely resemble that best available model, they are likely to be handicapped in one way or another. They may not live as long as those that do. They may not reproduce themselves as readily, and the trait that makes them different will either disappear entirely or continue in a recessive state, hidden in the genes, appearing only when two individuals that have the trait in common mate to produce offspring, some of which may reveal it again. Those offspring may not survive. But others that carry the trait recessively may.

That is why all of us lug around a great number of non-adaptive traits in our genes. Some of these act like policemen, stepping up to strike us down if we should deviate too markedly from the best model. Occasionally the handicaps are obvious; an exposed heart in a newborn infant is lethal, and the baby will die in a few minutes.

Most genetic variability is far less damaging and far less dramatic. It produces endless but very slight differences in all of us. But if it is potentially dangerous at all, it is reasonable to ask why it has persisted. Why doesn't every species gradually shake out the traits in it that are nonadaptive and produce only individuals that are most suitable for the life style of that species?

A moment's reflection will show that variability is

a necessity for all life. There are two reasons for this. The first is that the best available model is not necessarily the perfect model. There is always a certain amount of selection pressure within a species for self-improvement, for the development and intensification of traits that will fit it even better to its environment. Secondly, no environment is static. There must be genetic variability in a species if it is to change in response to the changes that are taking place all about it. Therefore, for all species, what may seem like genetic excess luggage today may turn out to be tomorrow's survival kit.

Primates illustrate very well these two competing forces in evolution: the tendency towards change and the tendency towards stability. Some of the ancestral insectivores evolved very little and very slowly, others more rapidly. The challenges that spurred that evolution may have been extremely subtle, as subtle as a slightly more intelligent or slightly stronger brother shrew—one that was a little better at catching the insect or attracting the female on the next branch. Over a long period of time, and in some places, "better" tree shrews resulted. In response to the evolutionary shaping that an arboreal way of life encouraged, they began to change rather rapidly. Jumping and clinging was a better way of getting safely and quickly about in the branches than the rat-like scuttling that had preceded it. The hind leg became longer. The front paw gradually lost its rat-like claws and acquired the kind of flat nails that are a hallmark of all primates today. All four paws began to turn into hands. The fingers grew longer and more flexible, and developed tactile pads at their ends. All these innovations greatly improved the ability of these new-model animals to move rapidly and sud-

denly in a tree—to grip a branch or to catch and hold a fast-moving insect or small lizard.

As leaping, clinging and catching became a way of life, dependence on smelling became less important than dependence on seeing, particularly for an animal that lived in a three-dimensional world of trees instead of the two-dimensional world of the flat ground and was being called on constantly to make precise judgements about how far away a branch or a lizard was. In response to the growing importance of seeing over smelling, the head of the ancestral tree shrew also began to change. Its snout became shorter, its skull rounder. Its eyes became larger and moved gradually towards the front of the head, where vision from one eye could overlap that from the other, giving the animal what is known as binocular—or stereoscopic—vision.

With binocular vision came a far greater ability to judge distances than is possessed by a creature whose eyes are located on the sides of its head, as a rabbit's are. Rabbits must be alert to what may be about to attack them from the side or from behind, but they have no need to see what they eat—grass does not move, and can easily be located by the nose. Nor does eating grass require a high degree of intelligence —less, certainly, than does hunting down elusive game in the tree-tops. In time the rounder skulls of the tree-top dwellers began to contain larger brains.

Within 10 or 20 million years those modifications had become sufficiently advanced so that a distinct new group of animals could be identified: the primates. The earliest examples are the prosimians (pre-monkeys), and, like the tree shrews, their descendants also survive. Among them are lemurs, lorises, tarsiers and bush babies. Some of them, particularly some of the larger lemurs, look and act very much like monkeys. If monkeys had never evolved, lemurs presumably would still exist in places where monkeys are now found.

Unfortunately for the lemurs, monkeys did evolve, establishing another fork in the primate family tree. At first, they could not be called anything more than late-model super-lemurs; the differences between them and the ordinary old-style members of the lemur populations were too small to be of much significance. But as these differences began to build up, through their survival advantage to the individuals that had them, ultimately the trees became filled with brighter, swifter, defter, altogether abler descendants that could finally be identified as monkeys. The lemurs faded away in most places because they could not stand the competition. Where they survive, as in Madagascar, it is because there are no monkeys there, and never have been.

So pencil in a fork—a pretty ancient one—to indicate the split between monkeys and prosimians. Then slide along the monkey branch to the next fork, which (on the tooth-cusp evidence previously given) marks the divergence of monkeys and apes. Then proceed on the ape branch to the next fork. At that point hominids will begin to appear. Once again, their appearance will have to be detected by clues supplied by their teeth and jaws.

When the jaw of a gorilla or chimpanzee is compared to that of a modern man, five differences are immediately apparent. First, the jaw itself is larger and heavier in proportion to the total skull size than it is in a man. Second, the teeth tend to form three sides of a rectangle, with a row of incisors across the front and all the other teeth facing each other in two parallel rows going towards the back. Third, the male's canines are longer than the other teeth; when the jaw is shut the upper canine projects down among the teeth of the lower jaw, and the lower canine projects upwards. Fourth, there are spaces between the teeth of the upper jaw to make room for those oversized canines. Fifth, the roof of the mouth, known as the hard palate, tends to be flat.

The human jaw has none of these characteristics. It is much smaller and lighter in comparison to overall skull size. The hard palate is arched, not flat. The teeth are all about the same length, with no oversized canines and, therefore, no gaps needed in the upper jaw. Instead of forming a rectangle, they form a curve, with the widest part of the curve at the very back of the mouth.

With these differences in mind, we can now return to the first Australopithecine skull found by Raymond Dart in South Africa, to see if it has any hominid characteristics. What made Dart so sure it was not the fossil of an ape or a baboon was the appearance of its teeth. No oversized canines, no gaps —just a nicely curved, human-looking set. What bothered others who examined the fossil or read about it was not the jaw but the rest of the head. It was tiny. That strange jaw was part of an ape face (no chin or forehead), backed up by an ape-sized brain. But, considering that this creature was estimated to be two million years old, more than twice as old as any other known hominid, it did not seem so remarkable to Dart or to his supporter, Robert Broom, that this peculiar mixture of ape and human characteristics should exist in a fossil. Two million years, they reasoned, might bring one pretty close to a common man-ape ancestor. That ancestor could well display a confusing and unexpected mingling of characteristics.

The situation was further confused when subsequent Australopithecine finds made by Broom and other South African workers made it appear increasingly likely that there were two kinds of erect man-apes in South Africa. Years later, when numerous fossils of both types had been recovered, a clear scientific distinction would be made between them. One would be called *Australopithecus robustus* out of respect to its larger size and weight of up to 150 pounds. The smaller 80-100 pounder would retain the name of *Australopithecus africanus* that had originally been given it by Dart.

There was something about the two types that disturbed Broom greatly. Although accurate dating continues to be impossible in South Africa, he came to the conclusion that some of his Robustus finds were possibly as much as a million years younger than Dart's Africanus. This would not have bothered him had not the larger, younger man-ape seemed the more primitive of the two. Its jaws and molars were massive, less like those of modern man than the jaw and molars of Africanus were.

Could the younger, more primitive type be the human ancestor? It just didn't make sense. That heavy jaw and those oversized grinding teeth suggested that their owner was a vegetarian, chewing up large quantities of green stuff, much as a gorilla does today. To reinforce this idea, Robustus had a ridge of bone running from front to rear on the top of its skull. This ridge is also present in a gorilla's skull, serving to anchor the large muscles needed for the heavy chewing that a gorilla's diet requires.

Assigning a rôle in human ancestry to this creature

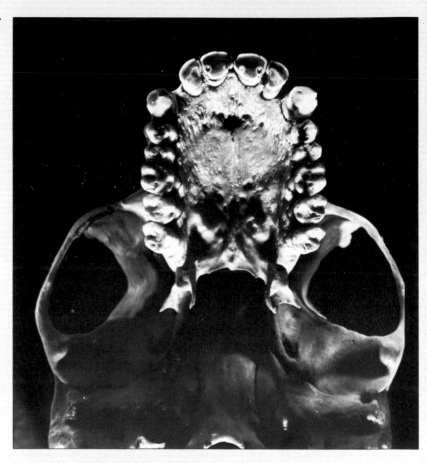

There is no mistaking the upper jaw of an ape (above) for that of a hominid (opposite). In an ape jaw—the example is from a modern pygmy chimp—the teeth are set in three sides of a box with the molars in nearly parallel lines. The four incisors across the front are separated from the other teeth to provide space for the large interlocking canines that all male apes have.

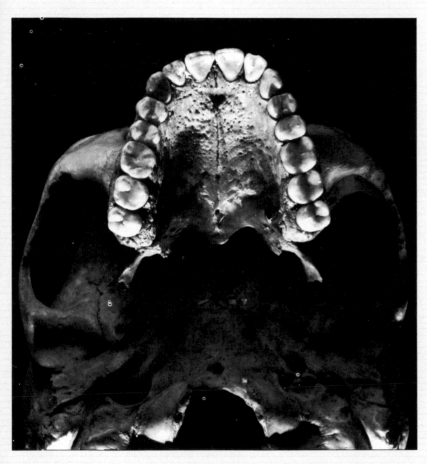

A human jaw is not rectangular but bow-shaped. The teeth run in a curve that is widest at the back of the mouth. Since human canines are small, there is no need for interlocking space in the jaw—all teeth touch. Human molars are larger in proportion to incisors than ape molars (opposite). The human jaw is also shorter: it does not project as far from the base of the skull.

raised awkward problems. One does not get specialized jaw equipment like this overnight. It can be assumed that Robustus had been following an evolutionary course towards a specialized vegetarian life for a long time. Therefore the most reasonable expectation would be for it to continue to do so—and not suddenly switch to the omnivorous (and smaller-jawed) way of life that man would come to lead a few hundred thousand years later.

Dilemmas of this sort are extremely troubling to palaeoanthropologists. Evolution does not work that capriciously or that fast. It is much more logical to assume that since man is known to have been an eater of all sorts of things for at least three-quarters of a million years, he probably has had that trait far longer. Broom hoped to throw light on this dilemma by finding some other clue to human-ness in one of the two South African fossil types. He and his colleagues hunted for years for stone tools that might be associated with either the robust of the smaller, gracile man-ape. For years he found none.

Then came that bombshell from the north: the Leakeys' discovery of a skull and tools in Olduvai Gorge. Disconcertingly, the Leakey find was that super-robust type he named *Zinjanthropus boisei.*

For a short time, that seemed to settle the argument; a very nonhuman creature seemed to be an immediate ancestor of man. Students of evolution had no choice but to chew on something that tasted very bad indeed. But, as has been the case so often in the study of ancient man, a bad taste can turn out to be only a bite from the unripe sour end of the fruit. Turn it around and bite again—with the help of new evidence or a new look at old evidence—and the flavour improves. In the case of Olduvai this happened with dramatic suddenness. Only a year after they had found their first skull the Leakeys found another. This one was also 1.75 million years old but was of a gracile type, and even more man-like than the gracile South African specimens. In fact, it seemed sufficiently human to separate it from the Australopithecines altogether. It was not, Leakey felt, a man-ape but a true human, and deserved to be classified in the genus *Homo*. Leakey christened this find *Homo habilis*, "handy" man, in honour of his being the tool user.

That Habilis, and not the super-robust Australopithecine, Boisei, was indeed the tool user has now been pretty well verified. The Leakeys subsequently collected from Olduvai a series of Habilis fragments indicating that this type lived there for more than half a million years—using much the same primitive tool culture the entire time, and slowly evolving until he closely resembled Homo erectus. The subsequent discovery of Erectus fossils, also at Olduvai and spanning a period from more than one million to less than half a million years ago, strongly suggested that one evolved into the other.

Habilis earned neither name nor credentials easily. He was primitive and small-brained. Many anthropologists preferred to identify him as an advanced type of the gracile Africanus and not deserving of *Homo* status at all. Many still so identify him. His qualifications as a distinct species have been in question from the day he was christened.

Skeletons of three Australopithecines are reconstructed here from the slim haul of fossils shown in colour. Africanus, the smallest, has no skull crest, suggesting human-like jaw muscles. He is more straight-legged, indicating good walking capability. Boisei, by contrast, has knees farther apart and bigger arm bones, indicating he may not have walked as erect. Robustus' leg bones hint that he walked fairly well.

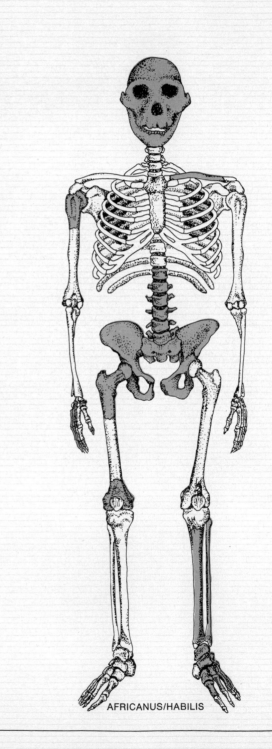

AFRICANUS/HABILIS

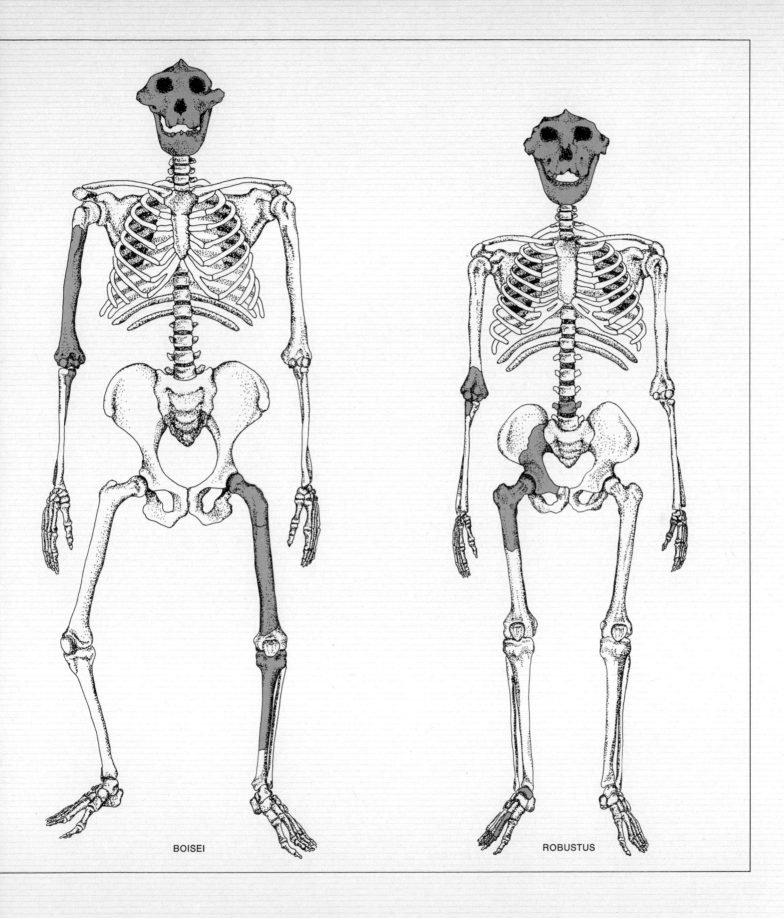

BOISEI

ROBUSTUS

Does Habilis mark another spot on the line where a fork is indicated? It all depends on how one looks at him. If he is an *offshoot* of Africanus (*i.e.*, if Africanus keeps on going), then a fork is indicated. If he is a *descendant* of Africanus—his man-like qualities becoming more recognizable over the course of time—then there is no fork. There is just a slow merging of one type into the other.

That second way of looking at Habilis seems to be the more sensible. But it still does not really clarify the matter of what Habilis actually is—whether he is a man or not. Compared to the certified human beings that came after him, he seems scarcely human. Compared to the more primitive types that preceded him, his human credentials suddenly improve. This disconcerting shift of perspective will always occur when the eye runs down a series of fossils that are related to one another through direct descent. The differences between them are differences in degree —not in kind—and will obviously become more pronounced as one comes forward in time. For hominids, some of the more obvious characteristics are: an increasingly large brain, a less pronounced bony ridge over the eye, a more delicate jaw and longer legs. But where does one draw the line?

The niggling query keeps coming up. But it is the wrong question. Since all creatures are bundles of characteristics, many of which may be evolving at different rates, drawing a line that is based on these characteristics will always cause trouble, as the following examples will show.

The British anatomist Sir Arthur Keith chose to draw the line at a point where the brain capacity touched 750 cubic centimetres. Anything below that, according to Keith, was not a man; anything above it

was a man, with Homo sapiens up in the 1,200-1,600 range. More recently another Britisher, Sir Wilfred Le Gros Clark, put the minimum at 700 cubic centimetres. Clark's choice, unlike Keith's was not an arbitrary one; it simply reflected the state of the fossil record at the time—there were no "human" skulls known to exist with cranial capacities of less than 700 cubic centimetres. Implicit in this situation, of course, was the possibility that a "man" with a slightly smaller brain might be discovered any day. What would one call him—and if a still smaller one showed up, what would one call *him*?

Habilis laid this problem right on the scientists' doorstep. The great difficulty in deciding whether he was a man or not lay in the fact that the "type specimen," the first one to be found and named by the Leakeys, had a brain capacity of about 657 cubic centimetres—just under the limit. Since then three other Habilis skulls have been measured by two experts, Phillip V. Tobias, an anatomist from South Africa, and Ralph Holloway, from Columbia University in New York. They came up with surprisingly uniform figures for these Habilis skulls, all of them from Olduvai. They range in capacity from 600 to 684 cubic centimetres, and can be averaged out at about 642 cubic centimetres. Too small-brained for a man? Certainly too large-brained for a typical South African gracile Australopithecine, whose mean cranial capacity is only about 450 cubic centimetres.

If brain size, or tooth shape or length of leg are all evolving confusingly at different rates, it is difficult to define species on the basis of these characteristics. Nevertheless, classification and naming are necessary. The best way to deal with this problem may be to assign a point in time, rather than a set of physical characteristics, to mark the emergence of a new species. Of course, it must be recognized, as one does this, that some blurring of characteristics along the way will always plague classifiers.

Where, then, does that leave Habilis? Dangling, it seems, somewhere on the edge of manhood. The debate about where to pigeon-hole him began with his discovery in 1960, and it was still creaking inconclusively along towards the end of the decade. Part of the trouble lay in the difficulty of "fitting" him according to his physical characteristics. But equally troublesome was the absence of anything to compare him to. There was only that solitary skull of a strange super-robust contemporary from Olduvai and a bunch of undated South African fossils. Looking back over Habilis' shoulder, deeper in time, was impossible. There simply was nothing known that was older than he was. Although Broom suspected that the South African fossils might be older, he could not prove it.

This lack of anything more than two million years old bore on another problem just as vexing: the coexistence of two kinds of Australopithecines in South Africa. Until they showed up, the prevailing opinion among scientists had always been that not more than one kind of erect human ancestor had ever lived during any one period on earth. There is only one kind of human today, and the presumption was that evolution and competition had always made them so. But as more and more fossils were recovered from South Africa, this belief was shaken. The robust and gracile types were sufficiently different to suggest that they represented two distinct species—unless, as some believe, the big ones were males and the small ones females of a single species. This is a fascinating idea that has never been laid to rest, although there

are some things wrong with it: for example, the distribution of fossils at the South African sites. If the theory is correct, then at some places the population seems to have consisted of nearly all males, at others it seems to have been almost all females. Inasmuch as all other evidence indicates that Australopithecine society consisted of bands of both males and females, this fact is hard to avoid.

To the theoretical question, *Could* two erect hominids have existed simultaneously?, South Africa does not supply a good answer. But if we go back to Olduvai and look at the two types there, a jolting surprise hits us. The robust specimen from Olduvai, Boisei, is so very robust that there is absolutely no question that it and Habilis, who is extremely gracile are different—although they coexisted.

The trouble with the big Olduvai skull was that there was only that one. Was it a single freak? Or did it represent a third hominid?

This was the state of the art in the middle 1960s: man's ancestry had been pushed back to about two million years, at which time a Homo who made tools and bore the somewhat shaky name of Habilis had turned up in East Africa. Although his discoverers and christeners, the Leakeys, did not (and still do not) think so, others concluded that he was probably descended from Australopithecus africanus. But until the dating problems of the latter could be solved, there could be no certainty about this. If older Habilis finds were made, or newer Africanus ones, there could be a real question as to whether Habilis was descended from Africanus at all. In that case there could be as many as four different kinds of hominids in Africa: a big one and a small one from Olduvai and a big one and a small one from the south—all un-

This list includes all Australopithecine fossils known up to 1971. Aside from teeth and jaws, the record is pitifully thin —there are only three hip bones and one shoulder blade. Furthermore, most of the bones are merely worn fragments. However, even these bits enable anthropologists to reconstruct the original shapes. The data for this list were supplied by Phillip V. Tobias and Bernard Campbell.

THE MEAGRE REMAINS OF AUSTRALOPITHECUS

SKULL	
CRANIA SKULLS WITHOUT JAWS BUT WITH SOME FACIAL BONES	11
CASTS OF SKULL INTERIOR	13
FACE FRAGMENTS	22
SKULLCAPS	45
UPPER JAWS	77
LOWER JAWS	79
BABY TEETH	110
PERMANENT TEETH	933

SKELETON	
SACRA (BASE OF SPINE)	1
RIBS (MOSTLY FRAGMENTS)	13
VERTEBRAE	20

HIP GIRDLE	
ISCHIA (BOTTOM OF PELVIS)	1
ILIA (HIPBONE OF PELVIS)	3
OSSA COXAE (SIDE BONE OF PELVIS)	9

SHOULDER GIRDLE	
SHOULDER BLADES	1
COLLARBONES	4

ARMS	
ULNAE (ONE OF TWO LOWER ARM BONES)	3
WRIST BONES	4
RADII (LOWER ARM BONES)	6
FINGER BONES (PALM)	9
HUMERI (UPPER ARM BONES)	10
FINGER BONES (DIGITS)	22

LEGS	
FIBULAE (OUTER LEG BONE)	1
ANKLE BONES	1
SHIN BONES	2
TOE BONES	5
ARCH BONES	12
THIGH BONES	14

related. What an anthropological mess that would be!

There was only one way to clean up that mess, and that was by getting more of the mud off the tapestry—by finding more fossils, by better dating of the ones that existed, by digging deeper into time. In the hope of doing this, an ambitious international expedition was organized in 1967 to look for hominid remains in Ethiopia. Its destination: a remote spot named Omo in the southern part of the country.

Omo had several attractions. For one thing, it had been visited 35 years before by one of the leaders of the new expedition, the French palaeontologist Dr. Camille Arambourg, and found to be rich in animal fossils. For another, it bore a striking resemblance to Olduvai. It, too, is part of the Rift Valley geological complex, a giant crack in the earth that runs south through Africa, its route marked by chains of lakes and rivers, its sides edged by towering escarpments. Much of the Rift Valley is dry now, its lakes shrunken, some of its cliffs worn away, its stones baking in the sun. There is no river at all today in the Olduvai Gorge except during flash floods, although the gorge itself was made by a river. There is nothing but rock, and heat—and fossils. Omo is hotter yet. Its river still runs down from the Ethiopian highlands and empties itself into Lake Rudolf, just over the border in northern Kenya. Lake Rudolf itself had grown and shrunk twice in the last four million years. It is still a sizeable lake 185 miles long, but it is only a fraction of its former self, and is currently shrinking. The brutal land around it is largely unexplored.

The Rift Valley is an uneasy spot on the earth's surface where large continental forces are still at work. It has long been a centre of volcanism. Cones and craters pock it. Thanks to neighbouring volcanic activity,

Olduvai has its invaluable layers of datable volcanic ash; so does Omo. Finally, both places were more hospitable in the past than they are now. Laced with rivers, much greener, supporting a far larger animal population, each provided the lush water-edge environment, the forest-becoming-savanna that early hominids are believed to have preferred.

But Omo is also different from Olduvai, and the differences are what made it particularly appealing to the new expedition. At Olduvai, the most accurate and useful pages of the volcanic timetable are crowded into a 200,000-year period that is not quite two million years old and not much over 100 feet thick. At Omo the strata being investigated are more than 2,000 feet thick and span a far longer period of time. Moreover they contain a great many volcanic ash layers. These come at varying intervals, some of them only 100,000 years apart, some more widely spaced, each datable by potassium-argon methods. Together, they can be used by scientists to step backwards into time, into the earth, layer by layer, as if they were descending the rungs of a ladder, noting the date on each rung as they go.

But one does not have to dig at Omo to go deeper into time. The strata have been heaved up in the past and now lie at an angle to the earth's surface. It is necessary only to walk along to find successively older layers revealing themselves like the bones of an immense buried skeleton, its ribs poking up one by one.

Best of all, the Omo succession picks up about where Olduvai leaves off—about two million years ago—and goes back from there. How far back this was had to be worked out by a team of geologists whose all-important job was to unscramble the rock history of the entire area so that everything could be properly dated. This has been done. The lowest fossiliferous layer turns out to be more than four million years old—twice the age of anything at Olduvai!

In this area of great promise, the members of the 1967 expedition settled down with high hopes. Arambourg picked the spot he had worked before and knew to be productive. A second group, under the direction of an American, Clark Howell of the University of California, went a short distance up the river Omo to tap a previously unexplored area. A third group, headed by Louis Leakey's son Richard, picked another untapped spot across the river from Howell. As it turned out, Richard Leakey's choice was the only one that proved unfruitful. There was plenty of material there, but his side of the river contained layers that were not old enough to be of interest to the expedition. He decided to disassociate himself from it and return to his home country, Kenya, and do some prospecting there—a decision that would prove to be one of the most fateful in palaeoanthropological history.

All the other members of the groups persevered where they were. Immediately they began recovering extinct animal fossils in a richness and variety unmatched almost anywhere. The great number of dated layers at Omo made it possible to trace the flowering and dying, the evolutionary changes that had taken place in some 80 species of mammals. Six genera and eight species of extinct pigs laid their secrets bare in the strata. Twenty-two different kinds of antelope were discovered, several extinct sabre-toothed cats. So varied was this haul and so complete a story did it tell that matching available fossils from other places with those in Omo became a distinct possibility. Perhaps a key exists after all to the age of those cryptic animals in South Africa that Broom classified and then regretfully laid aside years before because he had nothing to match them to. Turn that key, and a more reliable date for the South African hominids is provided!

In addition to all the useful pig and antelope fossils found at Omo, hominids began to appear. The French were first with a jaw. Teeth were found by both parties, eventually 150 of them, together with other jaw fragments, parts of two skulls, two arm and two leg bones. Not a spectacular haul, but an enormously significant one for a couple of reasons. First is the great age of five of the teeth: 3.7 million years. Four of these teeth are definitely of the super-robust type, the same kind as that one odd Olduvai skull that Louis Leakey named Boisei. But the super-robust fossils in Omo are twice as old as the Leakeys' find. Furthermore, other specimens continued to turn up in various layers right up to 1.8 million years ago. Boisei apparently lived at Omo for at least two million years.

Equally important is a group of 19 teeth and part of a thigh bone recovered by Howell from a stratum three million years ago. These are not Boisei teeth, but resemble the gracile fossils of South Africa. If further analysis should prove they are indeed of the gracile type, it will mean that a good date has at last —after nearly 50 years—been provided for that elusive little Australopithecine.

The Omo work continues, and the stream of geological information, and animal and hominid fossils, pours out of it unabated. Dr. Arambourg died in 1969. His place was taken by his long-time associate and compatriot Yves Coppens.

Meanwhile Richard Leakey had decided to do some further prospecting of his own. He buzzed off to the south in a helicopter, across the border into Kenya and along the eastern shore of Lake Rudolf. In one of those classic episodes in which an adventurous young man follows a hunch to a vast fortune, he spotted some likely sites from the air and set his helicopter down almost on top of what is beginning to turn out to be one of the richest mines of hominid fossils ever located.

The results of Richard Leakey's first years at East Rudolf are sensational: three superb skulls, more than two dozen mandibles or parts of mandibles, some arm- and leg-bone fragments and some isolated teeth. About two thirds of this material is of the super-robust Boisei type and extends through a time period that lasted from a little over two million years ago to about one million years ago. Adding this to the Omo finds, there is enough material here in the way of young and old individuals, males and females, enough variation in dentition, for the outlines of a variable population of super-robust Boisei Australopithecines to begin to reveal itself.

Having a population to study instead of an individual fossil is enormously important. No two people today are exactly alike; no two Australopithecines were either. It is for that reason that drawing conclusions from a single fossil is risky. Measurements taken of it, and theories spun off as a result of those measurements, may be misleading because the part being measured may not be typical. It is only when a large number of specimens is available that all their variations can be taken into account, and a norm derived from them. If a visitor from outer space were to describe and name Homo sapiens sapiens by examining one skeleton, that of a short, squat, heavy-boned New Guinea tribesman, he would certainly be excused if he set up another species on the basis of a second skeleton discovered later a few thousand miles away—that of a seven-foot, slender-boned Watutsi tribesman from central Africa.

That is why the Boisei population that is emerging is so valuable. It begins to reveal limits to variability beyond which none of its members go. Anything that does exceed those limits can be presumed to be something else. And those limits are now well enough defined to make it quite clear that the gracile Australopithecines *are* something else; there is absolutely no question about it.

Furthermore, the Boisei population seems to be different from the robust types of South Africa, which also exist in sufficient numbers to constitute a variable population with norms of its own. These norms do not overlap Boisei norms. It is now clearer than ever that Boisei seems to be Robustus gone super-robust. It, too, has a bony crest along the top of its skull, but a more pronounced one for the anchoring of even bigger muscles to work a more massive jaw containing larger molars. Even Boisei's premolars are on the way to becoming molars. All these are indications of a life adapted to the eating of large amounts of coarse vegetable matter. Apparently this way of life was not conducive to rapid evolution, since Boisei existed in East Africa for nearly three million years—possibly much longer—without changing much. Does a phlegmatic, vegetarian, perhaps non-tool-using existence account for this static evolutionary picture? At this point it is impossible to say. We can only note that this is the picture that Boisei gives us—stagnating comfortably in its own niche until challenged by another hominid.

Unfortunately for Boisei, there was another hominid, and the challenge came.

Richard Leakey's East Rudolf finds also include a series of late-gracile-into-Homo fossils. For a million years this type lived alongside Boisei; their coexistence is confirmed without question. At first their paths—their ways of making a living—did not cross. But as time went on this situation must have changed. Though Boisei did not evolve significantly, the gracile hominid did. Its brain became larger, and this apparently spelled disaster for all the robust Australopithecines, both north and south. Boisei disappeared in East Africa about a million years ago. According to Broom, Robustus disappeared in South Africa at about the same time. There are, in fact, no reliably dated robust specimens anywhere that are less than a million years old. The merging gracile-into-Homo, his niche expanding along with his expanding brain and his expanding capabilities as a hunter, apparently crowded his lumbering cousins off the face of the planet.

What is the gracile-into-Homo type that appears at East Rudolf? Richard Leakey suggests that it is similar to Habilis from Olduvai, and this has been seconded by others, notably Elwyn Simons and David Pilbeam, two specialists from Yale, who have studied photographs and casts of several specimens. Richard Leakey himself will not give a name to it. He is extremely cautious about labels. He simply calls all robust types Australopithecus and gracile types Homo, preferring to leave the more precise naming of species to others.

In this he is extremely wise. The entire Habilis story is not yet told. It is obvious that as the ancestry of Homo is worked farther back he will become less

and less of a Homo, and sooner or later will have to be called something else. In 1971 Richard Leakey found a jaw fragment that may be 2.6 million years old that he tentatively identifies as Homo. And in 1972 he announced discovery of a 2.5 million-year-old skull with a larger brain capacity than any Habilis specimen. But this brings Homo to within 400,000 years of those 19 gracile Australopithecine teeth found in Ethiopia. It makes him older than other gracile teeth found at Omo and dated at 1.8 million years. Now, at last, we find the gracile type and Homo overlapping in time. Since their physical characteristics also begin to overlap and mix confusingly, making it hard to say which is one and which is the other, this is probably one of those points in time where a cut in the chain is in order. In this case, following a suggestion by the British anthropologist Bernard Campbell, let us make the cut at two million years, and start looking for the missing link at about that date. Just what we will call it will have to remain unanswered for a bit longer.

Another of the fascinating surprises at East Rudolf is the presence of stone tools that are 850,000 years older than the oldest from Olduvai—and better made, presumably by those gracile types that

Richard Leakey calls Homo. Tools also show up at Omo, their age pinpointed at between 2.1 and 1.9 million years, thanks to that marvellous volcanic-ash calendar. Since most of the hominid fossils at Omo are super-robust, this raises for the second time the awkward question of whether or not Boisei was a toolmaker. It would be extremely neat if the Omo workers could resolve this problem as the Leakeys did at Olduvai, by finding Habilis remains to go with their tools. But unless—or until—they do, the jury must remain out on the matter.

Boisei may yet turn out to have been a user of tools, too, but a dim-brained one whose dependence on them was never very great because of his vegetarian dietary preference.

This limitation could explain why Boisei never evolved into a man. Increasing tool use is closely correlated with better manual dexterity and with gradual development of the brain. If Boisei had ever been interested in extending his diet to include much meat, his preoccupation with tools would probably have been greater, not only to chop sinewy and gristly parts of his prey apart, but also to hunt and kill it. But that rôle, that niche, was already being filled by another hominid. As a result, Boisei along with Ro-

The Lothagam jaw (far left, foreground is a 5.5-million-year-old piece of a lower jaw with one molar attached. Definitely hominid, it more closely resembles the human jaw, placed behind it for comparison, than it does any ape jaw. However, it is thicker —more "robust"—than the human jaw, which has been cut (near left) to show its thinner cross section.

bustus down in South Africa, certainly by two million years ago, and probably much earlier than that, was already doomed to disappear. For some millions of years he undoubtedly watched his smaller, livelier, increasingly bright, increasingly dangerous cousin scampering about, paying him no heed, oblivious of the fact that his eventual murderer was evolving right before his eyes.

While this scenario may settle the problem of where the small and the large Australopithecines went—the small becoming men, the large becoming extinct—it still leaves us with the same old question: where did they come from?

At this point the evidence becomes extremely sketchy. Small patches reveal themselves along the frayed bottom edge of the tapestry of hominid evolution, spaced so far apart that one can scarcely make out the pattern that connects them, their threads so worn that their forms are nearly illegible. There is part of an arm bone found by a Harvard University expedition at Kanapoi at the south end of Lake Rudolf in 1965. It is approximately four and a half million years old and is definitely hominid, with some suggestions that it belonged to a gracile type. Then there is a mandible found in 1967 at Lothagam, west of Rudolf, again by an expedition headed by Harvard University's Bryan Patterson. Its age: 5.5 million years. This appears to be the jaw of a gracile type. Finally there is a single molar, worn, but probably hominid, found at Ngorora, Kenya. Its age: an astounding nine million years. Behind these isolated and ancient fragments there is nothing: the Australopithecine line vanishes like a puff of smoke.

Or does it? If the dogged searcher, looking even deeper into the past, will now take another giant step backwards, he will encounter more jaw fragments and some more teeth. These are not from Africa at all, but were found in the Siwalik Hills of northern India. G. E. Pilgrim made the first find there in 1910, but did not recognize it as hominid. In 1932 the Siwaliks were visited again, this time by a graduate student from Yale, G. E. Lewis, who made another small find. The age of these fragments is uncertain. They may be 10 million years old; they may be 12. There is no convenient layer of volcanic ash in those hills for precise potassium-argon dating: as a result, age estimates have to be made by comparing other animal fossils present in these strata with similar fossils in other places. But the exact date of these Siwalik fragments is not as important as their nature. They suggest that their long-dead owners had short, stout jaws with curved sets of even teeth—much closer in appearance to those of the gracile Australopithecine than to the teeth of the apes that coexisted with them. These now-extinct early apes were the Dryopithecines. They existed in at least six different species. They were widely distributed throughout the warmer parts of the Old World. Some of them had already begun to suggest the gorillas and chimpanzees they would ultimately become.

Lewis's 1932 find was nothing like the Dryopithecine apes. He decided that it belonged in a different genus entirely, and named it *Ramapithecus*, after the Indian god Rama. Although Lewis lacked the extensive knowledge that we now have about Australopithecines, he still recognized his find as a hominid, and even went so far as to base his Ph.D. thesis, which he wrote in 1937, on his reasons for so thinking. But Lewis was young and obscure and the fossil jaw fragment was very small. His thesis was never published. It was not until the 1960s that Elwyn Simons, rummaging about in the fossil collections at Yale and other places, found some jaw fragments that had been given another name, compared them to Lewis' Ramapithecus and found that they matched. Ramapithecus now has some upper and lower teeth, and a better claim to human ancestry. On grounds of pure logic, it is tempting to regard Ramapithecus as a sort of proto-Australopithecine; after all, the Australopithecines had to start somewhere. But, however tempting such an idea may be, it is premature. We have no knowledge whatsoever of the nature of the rest of Ramapithecus' body. We do not know what its skull was shaped like or how large its brain was. We know nothing about its hand or foot. We do not know if it stood upright. All we know is that it was a widespread, and therefore presumably successful, animal, for Louis Leakey found another specimen 3,000 miles away under a layer of datable volcanic ash at Fort Ternan in Kenya.

Leakey's find is 14 million years old. He preferred to call it *Kenyapithecus*, but the consensus of present palaeontological thought is that it is simply another and somewhat more primitive Ramapithecus. The identification of these two fossils will be made more firm if and when more material is discovered. So far the East African and the India finds together have yielded a puny total of nine jaw fragments, some with teeth in place, in addition to a few loose teeth—not much to hang a theory on.

There, for the moment, the human fossil detective story ends. This is a most frustrating place to stop, for there are many unanswered questions that just a few key finds could clear up. The presence of Ra-

mapithecus in India, for example, suggests the possibility that man did not originate in Africa after all, that he may have been evolving in a number of places, and that some very old hominid fossils from Java—recently redated by potassium-argon methods and their ages recalculated to a possible age of 1.9 million instead of a half a million years—may be the descendants of a Ramapithecus line that never did set foot in Africa.

This could be so, but my own prejudice is that it is not. Leakey's Fort Ternan find locates Ramapithecus in East Africa at least two million years before it appeared in India—and, most important, in a country where Australopithecines of great age are beginning to turn up. It seems most sensible to continue to argue the case for Africa as the breeding ground of man. Ramapithecus populations may well have radiated out from Africa—to India and perhaps to other places—over a period of several million years. But this is no guarantee that they went on to produce hominid descendants in those places. Until fossil evidence confirms that they did, it seems more logical to assume that early hominid evolution was confined to the place where the fossils are: Africa.

Younger or better—or simply more—Ramapithecus finds could clarify this problem. They could also improve that species' claim to hominid ancestry, to say nothing of the light they might throw on the development of bipedalism. Without a Ramapithecus skull or key postcranial bone, there is no way of determining if it was an erect walker. We dearly need a gracile skull, a leg bone or a pelvis from Omo. These might shed some light on the timing of bipedalism there, perhaps on its rate of development, perhaps on how it developed; we are in the dark about all these things at the moment. A dramatic postcranial find of a super-robust type might reveal whether that animal ever was a true two-legged walker. Richard Leakey's recoveries from East Rudolf were the first to hint at a poor or only partially erect posture; the fragments of arm bones from there and the Omo are suspiciously heavy and long, and suggest that they were distinctively muscled.

The truth is, we need a full skeleton from one individual Australopithecine. The marvellous deductions about lines of descent that can be made from skulls, teeth and jaws have tended to obscure the fact that these parts do not say as much as they might about how their owners lived. Some interesting speculation can be made about those matters; it would be better to be sure. Nothing approaching an Australopithecine skeleton has ever been found. If all the known Australopithecine bones in the world were assembled to put together one skeleton, it would be a mock-up job from adult males, females and juveniles

—and many key parts of that skeleton would be lacking. The same is true for Habilis, that confusing late-Australopithecine character, and even for Homo erectus. The first complete set of bones belonging to one human ancestor are of a Neanderthal man, and they are about 60,000 years old. Here we are talking in a time range of one to fourteen million years ago. No wonder the fossils are so rare. No wonder they exist in such tantalizingly small fragments. The real wonder is that they tell us as much as they do. Accepting the Ramapithecus-ancestor argument, plus the other evidence offered here, one may conclude that:

1. Hominids have been separate from apes for at least 15 million years. 2. They had evolved into two or three different kinds by about five million years ago. 3. One kind continued to evolve, producing a better brain and a primitive culture. 4. These developments enabled him to exterminate his relative about a million years ago. 5. He has lived in solitary supremacy on earth ever since.

Chapter Three: Down from the Trees

Its brain small, its eyes angled sideways, the ring-tailed lemur survives as a living reminder of the extinct ancestor of all the primates.

The greatest events come to pass without any design; chance makes blunders good. . . . The important events of the world are not deliberately brought about; they occur.—George C. Lichtenberg

To hold an Australopithecine tooth in one's hand is a mind-boggling experience. Here is something that began in the womb of a female ancestor some three million years ago; something that followed the growth pattern of all modern human teeth, taking shape and hardening in the mouth of an individual Australopithecine infant that doubtless fretted when the tooth burst through its gum and made it sore; something that was then used for chewing during an Australopithecine childhood, and that ultimately was replaced by a permanent tooth.

This happened. The evidence is here in the palm of my hand. Hard, enduring evidence, stained brown by time, worn by years of grinding. This tooth chewed, crushed food for a tongue that tasted and that pushed it into a throat that swallowed. I stare at it. My thoughts go zooming off on a wild track, seeking to capture some shred of the individuality of the owner of that tooth. How did that owner live, how die? How was it regarded by its peers? As Australopithecine lives went, was its life a success or a failure? The tooth stares back at me. It says nothing.

It cannot. Fossils are very good at proving what the course of evolution was, less good on matters of how or why. This has been a source of frustration to experts for many years, so much so that some of them have begun to look for answers elsewhere. A field that seems fruitful to them is the study of man's closest living primate relatives, particularly the closest of all: the chimpanzee.

In addition to being practically a blood brother to man, the chimp is a good candidate for study for another reason. It is thought by scientists to be the least specialized of all the great apes, *i.e.*, the one that probably most closely resembles the ancestors from which all other apes—and man—have sprung. In other words, go back far enough along the hominid line and the ancestor we ultimately encounter may be not unlike a modern chimp.

Therefore, let us look at apes, but most particularly let us look at chimps to see if we can find, not only traces of ourselves persisting in them, but also a kind of blurred, distorted image of what we may have been like some millions of years ago. Equally important, we may discern a few of the ways in which we began to be different.

There is nothing new in the idea of studying primates to learn about human beings. As far back as the 1920s Robert Yerkes was observing domesticated chimpanzees in the United States and Solly Zuckerman was looking at Hamadryas baboons in the London Zoo. Both made important contributions to the field of primatology, but they and others gradually came to realize that if the intricacies of primate society were to be properly understood, the animals would have to be studied in the wild.

Unknown to most of the world, one man had already started this study. Eugene Marais, an obscure and eccentric South African poet, was observing the behaviour of baboons as early as 1905. But he had trouble supporting himself, and he eventually became addicted to drugs because of illness. Although he made some valuable contributions to man's knowledge of baboons (and also of termites) his work was largely ignored. Ironically, one person who did not ignore Marais was the Belgian Nobel Prize-winning writer Maurice Maeterlinck, who stole material from Marais and printed it as his own.

Organized field work on primates conducted by full-time scientists began in the 1930s, when C. R. Carpenter made pioneer studies of gibbons in Thailand and howler monkeys on Barro Colorado Island in the Panama Canal Zone. But the idea really caught hold after World War II, when a large number of young field workers, many of them destined to gain international reputations, began pouring out all over the world from universities and museums in a dozen countries. They studied gibbons and orang-utans in South-East Asia, langurs and patas monkeys in India, gorillas, chimpanzees, baboons and many forest monkeys in Africa. Studies of New World monkeys were started in Central and South America.

Primates turned out to be much harder to study than anyone had imagined. Many of them, like the mountain gorilla, live in inaccessible places. Many stay in the tops of jungle trees, where they are nearly invisible. Others, like the orang-utan, are extremely rare. Most are shy.

There is also the problem of what to look for and how to interpret it. Different species act differently in different areas, under different ecological influences, and even when their population densities vary.

Prejudices have had to be unlearned. For a hundred years or more, scientist and adventurer alike have cast the gorilla as a dangerous forest monster that beats its chest in rage and utters chilling roars of defiance, that has such enormous teeth and such stupendous muscles that it seems to have the option of whether to bite a man's arm off or simply tear it out by the roots. In the 1920s the explorer Carl Ake-

ley began to suspect that the gorilla was not all that ferocious. But it took a zoologist, George Schaller, to prove that the animal is actually shy and gentle.

Schaller's problem was not the danger of being savaged by gorillas but the difficulty of being close enough to get a look at them. He and his wife lived for 13 months in a small hut high on the flanks of the Virunga Volcanoes in the eastern Congo. There he ranged the fog-drenched forest, often climbing above 11,500 feet, walking 10 or 15 miles a day in unbelievably rough country, sometimes sleeping out, just to make contact with these elusive animals.

Schaller learned a great deal about gorillas but confesses there is much more yet to be learned. He never became truly intimate with a wild gorilla, nor did he ever touch one. That experience was reserved for Dian Fossey, a young primatologist who went where Schaller did, is there now and will remain for a number of years. She has come to know the gorillas better than Schaller did, and one day had the unforgettable experience of having a large male creep up and gently, shyly, touch her hand. To achieve even this fleeting contact she had to reassure the gorilla by keeping her eyes turned away.

Gorillas, though they are physiologically very closely related to human beings, are not much like humans in their basic life style. They eat coarse vegetation, are stolid, quiet, conservative, rather plodding animals. They do not remind us of ourselves nearly so much as do the more mercurial and venturesome chimpanzees.

Chimps, thanks to long and devoted field observations, are now well understood. One distinguished observer is Jane Goodall, who started her career as a secretary to Louis Leakey. Leakey knew of a troop of chimps that lived in a hilly forest tract near the Gombe Stream, a river running into Lake Tanganyika in western Tanzania. Since Leakey was interested in anything that had to do with primates, he wanted somebody to study the Gombe troop, particularly because he believed that the stream, with its forest and forest-edge environment, closely resembled that at Olduvai two million years ago.

Jane Goodall has since devoted many years to the Gombe chimps. Like Schaller, she ran into the problem of shyness. She set up camp near the lake shore, with her mother for company. Her days were spent roaming the forest, looking for chimpanzees in an area of about 15 square miles. Her plan was to watch the animals discreetly, not getting too close, just accustoming them to her presence as a preliminary to closer acquaintance. Months later she was still watching from a distance, still under suspicion. Ultimately, after a period of rejection that would have discouraged a less dedicated person, she was accepted, not by all the chimps but by many of them. A few she became extremely friendly with. In time she would spend thousands of hours with them, sometimes in actual physical contact, handing out bananas, playing with a baby, more often just sitting quietly and watching a society of unimagined subtlety and complexity gradually unfold.

When a photographer, Hugo van Lawick, came out after two years to take pictures of the animals, he had to undergo the same process of scrutiny, of familiarization and slow acceptance that Jane Goodall did, before the animals would act naturally in his presence. But, in testimony to the chimpanzees' intelligence, they associated him with their friend Jane and accepted him in only a month. Jane and Hugo are now married. Her studies and his pictures of the chimpanzees reveal an animal whose nature and social organization are provocative of all kinds of speculations about the emergence of man.

When we look at the great apes, our vision is clouded because we see them through the distorting glass of modern human eyes and in a setting that has become so overwhelmingly humanized that the apes tend to look a great deal more simple-minded, a great deal more vulnerable, a great deal less competent to get along than they actually are. The trouble is not so much with them as it is with the world. It has suddenly changed, too fast for them to change with it. It is overrun by a rushing, roaring, teeming, landscape-devouring, trigger-happy, forest-burning, polluting competitor. Today every species of ape is backed into a corner of the world, its habitat shrinking steadily before the onslaught of the lumberjack, the miner, the hunter, even the subdivider—for towns and cities now stand where apes once swung through the trees.

Nowhere is this remorseless process more vividly illustrated than in the Budongo Forest of Uganda, which I visited in 1968. This is a magnificent place, full of enormous hardwood trees, moist, green, tropical, so silent that the swish of the wings of giant hornbills can be plainly heard as they settle in the tree-tops—silent, that is, until one of the resident troops of chimpanzees lets loose with a chorus of ear-splitting shrieks, whoops, groans and yelps to announce to one and all that a fig tree full of ripe fruit has been found. Quickly the noise subsides again, and to a human observer crouched in the undergrowth, waiting for a glimpse of the chimps, the place might be uninhabited. When they are not moving and not shouting, chimps are remarkably quiet. I

Tree shrews, which now live in South-East Asia, closely resemble the small, insectivorous, rat-like animals that are believed to be ancestral to all the primates. They are true quadrupeds, with claws on their paws rather than the flat nails and fingers of other, more highly evolved primates.

remember lying under a bush one morning, hoping a nearby troop that had been boisterously yelling its head off a few minutes before would come my way when next it moved. The silence was profound. In this impenetrable green thicket it was almost impossible to believe that 20 or 30 large animals were going about their daily business within a hundred yards of me, eating, climbing, scratching. The only sound that came to me was a faint high whine, almost imperceptible. This was the noise of a sawmill a couple of miles away. Budongo is a state forest, and the Uganda government is steadily cutting it down for lumber. I lay there, wondering: can the chimps hear that noise or have they become so used to it that it means no more to them than the drone of insects?

There is a rustle in the leaves behind me. I try to turn my head without making a sound, and look straight into a rubbery face seamed with worried wrinkles. Bright brown eyes look back at me for a moment and are then withdrawn—so discreetly that I

am reminded of the Cheshire cat in *Alice in Wonderland*. There is no detectable movement, only a gradual disappearance, and I am left staring at a wall of leaves. The message of my presence is apparently communicated to the other chimps. Half an hour later I hear them hallooing from another direction.

Worried wrinkles? All chimps have these—or seem to. It is actually my 20th-century eye that has burdened the chimp with worry. Left to himself, freed of man, he has no need to worry in this forest. He belongs here. He is designed to prosper here if left alone. If I can bring myself to remember that, to look at him in that way, then he ceases to look vulnerable and pathetic. Take man back a few million years, divest him of all the things that now threaten the chimp, and the gulf between man and ape shrinks again. Men become less, apes more, particularly when the work of the Goodalls, the Fosseys and the Schallers reveal how subtle and complex primate societies are.

With that in mind, we can go back to a time before

The tarsier is a prosimian, one of the early primates that evolved from something like the tree shrews opposite. Tarsiers, now confined to South-East Asia, are tiny, nocturnal carnivores that eat insects and lizards. They have oversized kangaroo-like legs and four prehensile "hands", and their method of locomotion involves leaping and clinging.

there were any men, look at the primates as a group, try to understand why it was not a prosimian or a monkey but an ape—just one ape—that would follow a track that no other primate has ever followed. As a starter in this inquiry, we must first learn something about primate locomotion, for it was in the different ways that primates got about in the trees that the first clue to hominid evolution may be found.

The small rat-like insectivores that invaded the trees from the ground some 75 million years ago got around much as squirrels do today. But among those that would turn into bona fide primates, there was a rather rapid evolution. Paws turned into hands, with prehensile fingers for gripping. Some groups developed a way of getting about that depended on a slow but sure "four-handed" movement, characterized by an iron grip whose strength was out of all proportion to the size of the animal. Pottos still move this way, as do slow lorises, and both have powerful hands.

Another means of locomotion was in the direction of leaping and clinging. Some of the early prosimians were great jumpers. They had long legs—in proportion, as long as those of kangaroos—and very short arms. Most of them were extremely small. Some of the surviving species still are: the tarsier of the Philippines is no larger than a kitten.

But in time many of the prosimians grew. Exactly why has not been determined, except that there are forces at work in all species that encourage the selection of larger-sized individuals if there is no offsetting advantage in being smaller. For one thing, larger, more aggressive males have an advantage over smaller ones in competing for females. For another, an increase in size may protect an animal like a small prosimian from predation by certain small snakes or

hawks that are not powerful enough to take on a bigger one. In fact, a snake or hawk could encourage the evolution of larger animals in a population by killing a disproportionate number of the smaller ones and eliminating them as breeding stock.

But increased size brings problems with it. A large body is harder to hide than a small one. It also needs more food. If the food exists in the form of fruit or tender leaves and shoots at the tips of branches, a balance point will have to be reached where the advantage of being bigger no longer outweighs the disadvantage of being unable to get far enough out on the twig ends for the best food. In short, there is an optimum size for a particular way of making a living. If the selection pressure for larger size is great enough, then the way of making a living may change. A leaper may become a reacher, with longer arms.

Longer arms enable a primate to spread its weight among three or four branch ends, thereby reducing the weight on any single branch. Grasping hands with

flat nails become essential—claws will do for a very small climber but not for a bigger one.

Under these influences larger, heavier, longer-armed, more dextrous primates began to appear in the Oligocene about 40 million years ago. Their food preferences began to be different. Accordingly, they sorted themselves out in different parts of the forest, in different kinds of trees, even in different parts of the same trees. Some of these longer-armed primates became quadrupeds, running easily along the branches on all fours. These were the monkeys. Others, whose arms became still longer, tended more in the direction of reaching, climbing, hanging, swinging. They were the apes.

This difference between quadrupedalism and brachiation (as the swinging movement of the apes is called) may seem insignificant. Actually it is profoundly important. While monkeys often sit fairly upright, and while some of them occasionally even stand on their hind legs, they are not really upright

The potto is a nocturnal African prosimian with woolly fur. It clambers very slowly and deliberately along branches and is enormously strong for its size, with a vice-like grip and an opposable thumb. It retains on each foot a "toilet claw" for grooming, but all its other claws have become flat nails.

Monkeys like this vervet are extremely adaptable animals, bigger-brained and more numerous and widespread than the prosimians, from which all monkeys descended. The vervet is a common arboreal African species. It has good manual dexterity, and flat nails on all fingers and toes.

Of all the apes, the gibbon is the least closely related to man.
It is also the most specialized, with long powerful arms and
hook-like hands for swinging from branch to branch. Gibbons
live in dense South-East Asian forests. In that narrow world
they stick to one-family territories. Perhaps because
they seldom come to the ground, and thus miss the stimulus
of large-group activity, they are the least intelligent apes.

animals. They get down on all fours to move about. Also, while they do have well-articulated fingers and toes, they walk along the branches with the palms of their hands and feet. Thus, although they pick up things with great dexterity when they are sitting still, they must drop them when they run—their quadruped mode of locomotion compels it.

Apes, because of their habit of reaching, climbing and swinging in trees, are essentially more erect animals than monkeys, and have greater freedom of arm movement.

The greater the degree of erectness and arm flexibility, the more sitting, standing and reaching an animal can do, and the greater the emphasis that can be put on hand use. The hand becomes increasingly free to grasp, to pluck, to hold, to examine, to carry. The more the hand is used for these acts, the better it gets at doing them. A chimpanzee, as Jane Goodall discovered, has the manual dexterity to strip the leaves from a twig—in other words, to make an implement—then deftly insert that twig into a small hole in a termite mound so that it can lick off the termites that are clinging to the twig when it is pulled out. This remarkable act not only requires precise manipulation of a rather high order, but it also requires the intelligence to do it, which is another way of saying that increased dependence on hands has an evolutionary effect on the brain: it gets bigger.

Proof of this is that apes—creatures whose hands are freed by their potential for erect posture—are, as a group, conspicuously more intelligent than monkeys, whose hands are dextrous enough but whose quadruped way of life limits their use, and thus limits the stimulation that hand use has on the brain.

Fair enough. Apes have the potential to be erect an-

imals. They are cleverer than monkeys and they use their hands more. Why didn't all of them become men? This is an extremely complicated question. Perhaps the best way to answer it is to put oneself back in the equatorial forest, in a Budongo of 20 or 30 million years ago, and try to visualize the situation from the point of view of the apes of the time. There were already differences among them—we know that; their fossils tell us so. But those differences were not so great as they are now. All four of the surviving ape species—gibbon, orang-utan, chimpanzee and gorilla—are larger than they once were, all but the gibbon much larger. All of them are longer-armed, the gibbon and orang-utan considerably so.

The gibbon and orang are Asiatic species. By a great number of external and internal measurements, including genetic ones, they turn out to be remarkably different from chimps and gorillas. In fact they are much less like a chimp than a man is. This indicates a separation far back in time, long before the man-gorilla-chimp separation, and even suggests another argument for the common African origin of the latter three, more closely related species.

A point worth noting about the gibbon and orang is that both are largely arboreal animals today. Millions of years of climbing and swinging and a total reliance on the fruits that grow in jungle trees have brought them to an extreme pitch of arboreal specialization. When they do come to the ground they move slowly and uncertainly. In the trees, however, they are superb, each in its own way. The gibbon is an airy flier that hangs from branches, swinging from one to another like a pendulum loosed from its fitting and suddenly hurtling through space in a breathtaking arc, grabbing the next branch just long enough to launch itself in the direction of a third. Any gibbon could effortlessly travel the length of an underground carriage in a couple of seconds with two or three swooping swings, using the straps that hang from the ceiling. With this animal, arms and hands are everything. Its fingers are extremely elongated, specialized to serve as powerful hooks to catch branches. As a result of this finger specialization the gibbon has the poorest manual dexterity of any ape —and the smallest brain.

The orang is quite different from the gibbon. It is much larger; adult males weigh over 350 pounds, compared to the gibbon's 10 or 15 pounds. Obviously an animal of this size cannot go careering through the branches. But thanks to the development of four extremely prehensile hands, and limbs that are so well articulated that they can reach in any direction —in a sense, four arms—there is almost nowhere in a tree that an orang cannot safely go, despite its great bulk, by careful gripping and climbing. It moves slowly but confidently, and gets into some bizarre attitudes. An orang may grasp a couple of vertical boughs with its feet, its body hanging in between but in an upright position so that its feet are above its shoulder, while it reaches out with an arm for fruit. Orangs spread-eagled in trees bear an uncanny resemblance to huge, hairy, orange-brown spiders.

When these specialized animals are compared to the general-purpose monkey-ape model that was probably ancestral to all, it is clear that the gibbon and orang have evolved in a direction different from the one that might have made them human. Each is far too specialized for tree life to become anything but a more extreme example of what it is now.

Gorillas and chimps, on the other hand, have not travelled the exclusively arboreal route. Whatever specialization that has taken place in the gorilla has been in the direction of a great increase in size, along with a dietary switch from the fruit and leaves found in trees to a more generalized menu of fresh bark, larger leaves, roots, bamboo shoots and other plants.

These two specializations of the gorilla go together. In coming down to the ground to eat the things it does, the gorilla can best take care of itself there by being so large and strong that other animals cannot attack it. Because it is so large it needs a great deal of just the kind of coarse greenstuff that it finds in large quantities in the places it inhabits. Today it might be called an ex-brachiator. It retains the equipment for climbing and reaching, the deft hands, the good brain, the long arms of a brachiating ape, but it is too bulky to brachiate. Young gorillas are frisky and venturesome in the trees, but their elders are not. They are essentially ground animals. Having carved out a highly successful niche for themselves there, they are under no evolutionary pressure to evolve further. They are, in a sense, the elephants of the primate world. And, like elephants, they are fortresses in themselves —secure against anything but man.

Of all the great apes, the chimpanzee is the least specialized. In size it is a neat compromise—not too big to go about in trees, big enough to take care of itself against ground predators, particularly since it travels in troops that are collectively formidable. As a result, it is at home in both worlds. Although the chimp is still a fruit eater, whose favourite staple is ripe figs, it will also eat a wide variety of other things that it finds on the ground, including meat: birds' eggs or fledglings, insects, lizards or small snakes, occasionally a young baboon, bush pig or bushbuck.

Whether the chimpanzee is innately more intelligent than the gorilla is hard to say because the latter has been so little studied—but from what is known, the edge would seem to go to the chimp. Its nature certainly makes it *seem* brighter. Chimps are sociable, curious, extroverted. They like to please. This characteristic may be deeply rooted in a pattern of social behaviour that they have worked out. Living in groups, they must find ways of defusing frequent and potentially dangerous confrontations with one another. They do this by reassuring, by touching and appeasing. "I'm a nice fellow," they seem to be saying. "Look at me, watch me; you'll believe it."

Chimps are also copycats, and very observant ones. Their open, gregarious, let's-try-anything society encourages this. In addition to using twigs as tools, chimps use rocks to smash things. They throw sticks and stones and brandish large branches in response to threats. They use crunched-up handfuls of grass or leaves as sponges to hold water. They use the flanged buttress roots of forest trees as drums, beating on them with the palms of their hands.

Gorillas, by contrast, are slow, introverted animals that seem to prefer things simple and understated. They seldom fight. It is almost as if their tremendous physical prowess has been deliberately muted by a personality that emphasizes self-centredness, tolerance and a kind of brooding introspection, all in order to prevent them from doing terrible damage to one another. Except for throwing leaves and branches when they are disturbed, gorillas are not known to use tools. Young ones play with things, as do the young of many mammals, but adults never do. They are too phlegmatic and humourless for anything like that.

The behaviour of chimps and gorillas has probably evolved no faster than their physical evolution, and we may assume that something like the present societies of gorillas and chimps has existed for some millions of years. That the stolid gorilla got side-tracked into being an unenterprising vegetarian is not hard to understand. Harder, at first glance, is to see why the chimp—whose attributes seem to be just the kind that another tree-descending ape changed into manhood—did not become a man too.

On close examination there really is nothing mysterious about the failure of the chimpanzee to rise above its present status. The modern chimp represents the workings of the familiar process of speciation, the gradual separating out of a single population into sub-populations, each aimed in a slightly different direction towards the eventual occupation of a different niche. Assume that the process starts with an unspecialized, all-purpose ape, not too different from a chimpanzee. It is probably somewhat smaller, somewhat shorter-armed, somewhat more catholic in its food tastes than the chimp of today, and hence more willing to look anywhere for things to eat. The world, in short, is wide open to it. It can go in any one of several directions.

If, in one part of that ape's range, there are large forests and an abundance of fig trees, the temptation to stay in the trees and become increasingly specialized as a fruit eater and brachiator will be very great. I say "temptation". I do not mean to imply that there would ever be any choice as far as the apes were concerned. They would live out their lives, generation after generation, blindly doing what came easiest.

In another place or at another time, however, the environment might be somewhat different: fewer fig trees, but an abundance of seeds, berries, tubers, insects and other food materials on the ground. This environment could encourage the development of an animal of a slightly different shape and habits. Tree-dwelling brachiators depend primarily on their arms for movement. Ground-dwellers need stronger legs to walk. If they have a long history of sitting, standing erect and hanging in trees, they can be expected to do a good deal of standing up on the ground—perhaps to look over tall grass as some monkeys and apes now do, perhaps just to move about.

Since these ground-orientated apes already have hands that are well developed for holding things, there will be an added incentive to get up on their hind legs, for that will be the easiest and most convenient way to carry food. If they also have the beginnings of a tradition of using rocks, twigs and branches (and we know that chimps do), they will probably carry those around with them also, further enhancing the selective pressures for spending more time on their hind legs. Could this conceivably lead to the evolution of an erect, large-brained, sophisticated, culture-orientated chimp?

Theoretically it could—if we were willing to call the erect descendants of that pre-chimp a chimp. But we don't do that. We call the erect one an Australopithecine and we call the fig eater that stayed in the trees a chimp.

Proto-chimp and proto-Australopithecine might even have shared their environment as they underwent the slow process of evolutionary change, gradually becoming separated from each other by behavioural differences as well as by geographical barriers. We will never know what it was that first sent them travelling in different evolutionary directions. It may simply have been that the best fig pickers hogged the figs, unwittingly encouraging the development of a strain of apes that, willy-nilly, found it easier to survive on the ground than in competition with those big fellows in the trees.

Again, these things do not happen overnight. They take place with agonizing slowness, and totally without the awareness or consent of those in whom the changes are taking place. No ape ever "decided" that its future lay on the ground and that henceforth it would not associate with its tree-dwelling friends. What happened was that over an immense amount of time there developed a strain of ape that *habitually* looked to the ground for its sustenance, in the process acquiring a set of physical and behavioural characteristics that fitted it very well for such a life. The chimp, on the other hand, remained—and it still remains—a long-armed fig eater. It never truly abandoned the trees; it never had to.

To summarize: the critical matter here seems to be timing. You cannot leave the trees too early—as a quadruped monkey—or you will remain a quadruped, like a baboon. You must wait a while, until you become a reacher and climber and until the potential for erect posture and hand use is well developed in you. But you cannot wait too long or you will become too much of a brachiator (see what happened to the gibbon and orang; they are now stuck in the trees forever, with their preposterously long arms). You have to time it just right, not get yourself wedged too tightly in any particular niche, then be the *first* ape to do the "new thing" on the ground. That way, you get squatter's rights. That way, you become a man. The others, who might have become men, hesitated a little too long, then went in slightly different directions—and became a chimpanzee and a gorilla.

Where the Fossils Are— the Famous Hominid Sites of East Africa

In search of man's past, Richard Leakey's expedition negotiates patches of dried salt in the Chalbi Desert en route to Lake Rudolf.

Man's hominid ancestors evolved in a narrow north-south region of forest-bordered lakes and rivers that runs 5,000 miles from the Middle East to South Africa (*map, page 13*). This is the famous Rift Valley system, a place of geological uneasiness that for more than 20 million years has acted like an immense expansion joint, widening and deepening as sections of the earth's crust have shifted.

These fluctuations, coupled with climatic changes, have periodically raised and lowered water levels in the Rift. The run-off from lakes and rivers has not flowed to the sea, as it does elsewhere, but has been contained in an inland depression like the Great Basin of the American West.

The effect of containing this water has been threefold. Firstly, for a very long time the region provided an environment favourable to the evolution of man. Secondly, it provided an environment favourable to the preservation of fossils: since water cannot reach the sea, silt slowly fills the lakes; bones tend to remain where they fall, gradually being covered with sediments or volcanic ash. Thirdly, recent erosion has laid bare old sediments in spots, making it possible for the bone hunters to do their work.

The Badlands near Lake Rudolf

Some 20 million years old, the Turkana Grits is too old to yield fossils of hominids, which probably evolved no more than 15 million years ag

Aerial photographs of the Lake Rudolf area show why the map of this part of the world is a blank: It is too desolate to support much life today—and experienced fossil hunters recognize that most of the region could not have supported early hominids either.

Knowing where not to look is as important as identifying a rich site, and the palaeontologist needs not only an explorer's hardiness but a geologist's imaginative eye. If he knows that the accessible rock layers, like those at left below, were formed before hominids evolved, then there is no point in looking there. And if the visible strata are as recent as those below, the earth that Australopithecus would have lived on lies too deep for excavation to be practical.

To pass quickly over such barren formations and find outcrops of promising age (*overleaf*), Richard Leakey has often used a helicopter. Ironically, on his first aerial survey in this area, he found a marvellously rich spot that he has been unable to locate since.

ese sediments came from Ethiopian rivers and covered the strata of Australopithecus' times. If he lived here, his fossils lie beyond easy reach.

A rare desert rose blooms after rain in the fossil-rich sediment near Lake Rudolf.   *On camel back, Richard Leakey (centre) and assista*

Richard Leakey's continued explorations east of Lake Rudolf—often made on camel-back because of the rough terrain—revealed that here were large areas promising to yield exciting finds of hominid fossils. The surface strata were the right age, ranging from 4.5 million to just less than one million years—the same period that saw man emerge from man-apes in this part of Africa. And it seemed possible that the bones of lakeside hominids and the herds of animals that lived alongside them had simply been covered with silt and clay.

Near the lake itself, at a place called Koobi Fora, the promise became a reality. On one search Richard Leakey came across a 2.6-million-year-old hominid skull lying fully exposed, waiting to be picked up, on a rocky outcrop. And there was much more to come. Fanning out north to the Ethiopian border was a vast site of about a thousand square miles where hominid fossils and tools have since been found in large numbers.

...oss otherwise inaccessible terrain near Lake Rudolf. From camels they were able to spot hominid fossils they could not have seen from a plane.

Lothagam: Source of the Oldest Remains

This flat avenue is the bed of a sand river cut through the Lothagam Hill deposits. Enough rain water flows here to nourish a few palm trees. A ve

Lothagam, near the south end of Lake Rudolf, is a small river gorge cut through old lake deposits. Here the deposits have been thrust upwards by earth movements to form a conspicuous bump: Lothagam Hill. The contents of the hill—layer on layer of lake-bottom sediments—now lie at a slant, jutting out of the surrounding countryside and revealing deeply-lying fossil-bearing strata older than any other Australopithecine sites. At Lothagam, fossil-bearing sediments that are between five and six million years old protrude from the ground. Recent erosion has been working away at the outcropping edges of those tilted layers, and it was in one of them that Bryan Patterson's 1967 expedition found the 5.5-million-year-old Lothagam jaw, the oldest Australopithecine fossil yet discovered. Just what specific label the Lothagam jaw should bear is unsettled. Some think it is gracile, others suspect it may be Boisei, but its condition makes precise identification almost impossible.

...arly hominid fossil, the Lothagam jaw, was found nearby in a sedimentary layer that overlies the reddish ones visible in the cliffs to right and left.

Omo: A Land Green as in Ancient Days

The river Omo banks are densely forested, as much of the area was when Australopithecus lived. His fossils were not found here but in nearby

The only place in Australopithecus' homeland whose environment is now close to that in which he lived is the banks of the river Omo (*below*), north of Lake Rudolf. The Omo brings rainfall from a range of high mountains not far to the north in Ethiopia—and is one reason why the lake has not completely dried up during the last few thousand years of extremely arid climate in the region.

The Omo today is a leisurely stream. It loops idly across a flat plain that was once buried beneath the waters of a larger Rudolf. Its colour comes from the load of mud and clay it carries, which it will drop into the lake to form yet another layer of sediment on the bottom. It flows reliably all year and, as a result, is able to support a rich belt of forest edged by savanna on each bank. Australopithecines, living today, would probably feel at home on the banks of the Omo —but hunters after their fossils must leave the river to search the eroded, barren areas surrounding the oasis.

...arren areas where the ancient sediments are exposed.　　*The fossil-bearing Shungura formation, near the river Omo, is deeply cut by erosion.*

Olduvai Gorge: Site That Turned Up the Missing Link

Olduvai Gorge is forked. One branch is in the foreground and the other runs behind the butte whose layered deposits are visible. The Leakeys' firs

Olduvai Gorge, some 400 miles south of Lake Rudolf, was once a lake whose bottom, two million years ago, lay where the deepest part of the gorge now is. Subsequently the lake slowly filled with layers of sediment brought down from the nearby mountains until its bottom had risen some 300 feet to form the flat plain visible on the horizon in the picture below. Then, about 50,000 years ago, water from seasonal rains began cutting into the plain, slowly forming the gorge as it sliced back down through the deposits of time.

It is in this gorge that Louis and Mary Leakey—spurred on by the discovery of stone tools—did their exploring. They investigated it intensively for five miles or more in both directions from the point shown here. In the process they not only found the first fossil evidence of Habilis, the 1.75-million-year-old direct link to man, but also pinpointed about 70 sites where tools, hominid remains or animal fossils—or all three—lie.

ominid find was near the very bottom of the gorge. More recent deposits, forming the canyon rim opposite, are about half a million years old.

Chapter Four: A Niche on the Savanna

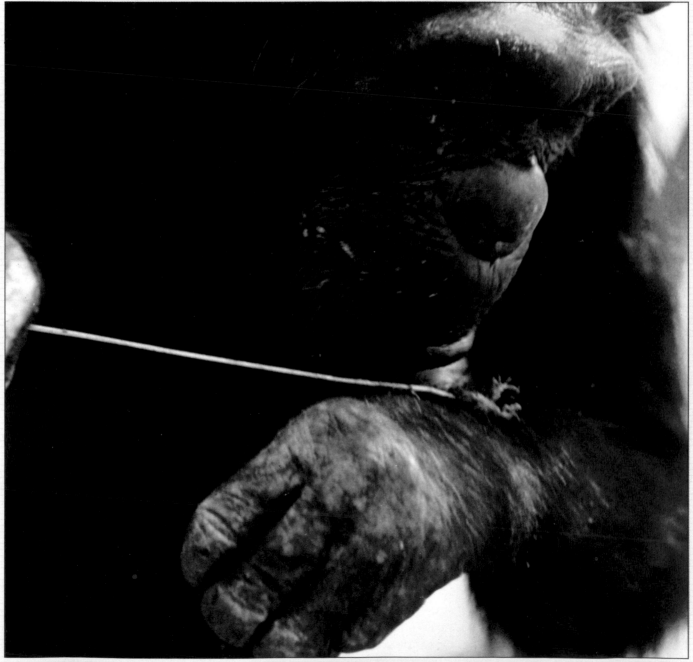

Like man's predecessor a practiced tool-user, a chimpanzee prepares to eat termites off the long straw it used to catch the insects.

Definition of a grapefruit: A lemon that had a chance and took advantage of it.—Oscar Wilde

In any successful detective story the clues must be presented in a way that gives the reader, as he encounters them, a growing awareness of the overall shape of the plot. They may be unexpected and surprising—in the best detective stories they always are —but they cannot be irrational, or they will simply bewilder. The trouble with the detective story of human evolution is that, although we know in a broad sense how the plot turns out (apes become men), we still know very little about the twists and turns of the story in between. There is an overall rationality to all the clues as they unfold—that we know—but we cannot be certain what it is.

To be sure, we have already learned some vitally important things. We know that Australopithecines existed, and that one kind evolved into man. We know something about the period of time involved in the process, and we have begun to get some insights into why the human estate came to be conferred on one particular primate and on no other. To learn that, we had to learn something about primates generally, their evolutionary history and the differences between apes and monkeys, in order to understand that there was something special about apes that made it possible for one of them to become a man.

It is now appropriate to look at some other clues in an effort to learn something about how that ape-into-man process may have taken place. Again, clues exist in two areas: in fossils and in primate behaviour.

"Do the new thing on the ground," I said at the end of the previous chapter. What does that mean? Did that first adventurous ape shin down a tree

and saunter off? Certainly not. He was surely more tree-bound than ground-bound for a long time while he and his descendants were gradually discovering that there was a living to be made on the forest edge and out in the open.

If he sauntered at all he did not saunter far. And how he sauntered is not yet known. Sherwood Washburn of the University of California at Berkeley, looking at other apes, maintains that early hominids were knuckle-walkers. Two of the great apes are knuckle-walkers today, he points out: the chimpanzee and gorilla. Their very long arms and short legs enable them to stand with their backs at a slant with the ground, their arms taking some of their weight, balancing themselves easily on their bent knuckles. As Washburn has noted, this is exactly the position a lineman (forward) in American football adopts before the ball is passed back or what a man may do when leaning over a desk or table. To get from there to an erect position is easy. Chimps do it all the time.

Since hominids were not quadrupeds when they came out of the trees, but brachiators of a sort (like the gorilla, they might be called ex-brachiators), it is logical to assume that they did what the other brachiators did when *they* came out of the trees. Knuckle-walking can be regarded as a kind of halfway step between quadrupedalism and bipedalism.

But it is not so regarded by everybody. The anthropologist Charles Oxnard has pointed out that the human shoulder blade resembles that of the orangutan, which is not a true knuckle-walker. This suggests that the ancestral hominid was large-bodied like the orang, was quite an accomplished hanger and swinger while in the trees and might have walked bipedally immediately when it descended to the ground (as that other swinger, the gibbon, does now) without ever going through a knuckle-walking stage.

Yale University's David Pilbeam adds an interesting observation about the backbone. The best specimen yet recovered of an Australopithecine spinal column has six lumbar vertebrae. Chimps and gorillas have only three or four, which suggests that they may have lost a couple in the course of their evolution as knuckle-walkers. A hominid that still has five or six lumbar vertebrae therefore may never have gone through the knuckle-walking stage.

Whether our ancestors pushed themselves upright from their knuckles or staggered inefficiently on two legs from the beginning, they did not become good walkers overnight. It is far easier to visualize a gradual shift to effective bipedalism over a long period of ground existence, during which selective forces encouraged an erect life, than it is to visualize a sudden one. For one thing, efficient biped walking requires a kind of foot and pelvis, and the development of leg and buttock muscles, that arboreal apes do not have.

What kind of forces are we talking about? Again Washburn has an answer. He reminds us that apes, because they are brachiators, have a tendency towards erectness that monkeys lack; some are also tool users. He places the ape-that-will-become-a-man on the ground, in a new environment where there are things to be picked up, rocks to be thrown, branches and clubs to be waved in threat displays or in defence. He presumes a gradual shift in eating habits from reliance mainly on the fruit found in trees to a diet of all kinds of things found on the ground. Items in this new diet need to be broken up, mashed, killed, fought over, fought *for* in competition with other animals. He puts all these factors together, finds this ground-dwelling ape using its hands more and more to carry objects, to work with objects, to fight with objects —and he comes to the conclusion that all this use of objects was the propelling force that drove man's ancestor permanently upright.

In short, man became biped because he became a tool user. As a bit of fossil evidence to back up early tool use, Washburn calls attention to the extreme smallness of the male Australopithecine canine tooth. In all other large ground-dwelling primates —chimpanzee, gorilla and particularly baboon—the male's canine is an enormous tooth, a true fang. One of its uses can be presumed to be self-defence against the large and dangerous predators ground-dwellers are exposed to. For a male hominid to be able to get along on the ground without such a tooth, we must logically supply some other means of self-defence. Washburn's answer: tools and weapons.

The argument that tool use spurred bipedalism can be turned around, of course, as several experts have pointed out, notably Bernard Campbell, the British anthropologist, and J. T. Robinson of South Africa, now teaching at the University of Wisconsin. The reverse argument assumes that man was bipedal from the time he first stepped away from the trees, and that it was this characteristic that gave him the opportunity to become a tool user by freeing his hands to carry things. If a hominid walked on two legs from the beginning, the argument goes on, then natural selection would inevitably improve whatever pelvis, leg bone, foot bone and musculature the hominid had, to make it easier for him to get about in that way.

There may be disagreement as to whether tool use first stimulated walking or walking stimulated tool use, but there is absolutely no disagreement on the importance of tool use in stimulating brain development and urging hominids along the path to manhood.

It is worth noting that while chimpanzees are tool users and toolmakers, they do not *need* tools. They can get along well without them. Nevertheless there is that extraordinary trait that chimps have—largely undeveloped, not vital to survival, but there. If nothing else, the chimpanzee proves that an ape with good manual dexterity can develop a tradition of elementary tool use simply by spending time on the ground where there are objects to be picked up.

What chimp first learned to fish for termites with straws, or how long this took to become an established activity, will never be known. Jane Goodall is not yet sure just how the present-day chimps do it —whether they have the innate intelligence to conceive of such an activity all on their own, or whether it is something that each new generation learns from the previous one. She does know that the youngsters have an opportunity to learn from their elders and that they do watch them intently, often copying what they do.

Another talent Jane Goodall observed among the Gombe chimps was for throwing things. Her accounts of this are extremely interesting for several reasons. For one, they reveal that the trait was well established; a number of individuals tried it, and under a variety of circumstances. For another, they suggest that the chimps' throwing, even though often woefully inaccurate, was actually useful. In chimp life there is a great deal of bluffing and aggressive display. During such activity an animal will jump up and down, wave its arms, hoot, shriek, charge forwards. This behaviour is even more disconcerting if it is accompanied by a shower of sticks or rubbish or stones. The fact that it is useful explains, of course, why it is now established as part of the behavioural equipment of the species.

At its present stage of throwing development, the chimp is scarcely a star athlete. Through lack of practice it cannot throw far, nor can it count on hitting anything more than four or five feet away—nothing to impress a human talent scout. But we must remember that the performance does not have to impress humans, only other chimps, baboons, leopards and the like. For that audience it is extremely effective. Throwing clearly does have selective value, and we can speculate that if chimps are left to themselves long enough they might become better throwers than they are.

At present, potential for improvement exists among certain individuals. There was one such at the Gombe Stream named Mr. Worzle (Jane Goodall named all the chimps as fast as she could recognize them, for ease of identification). Mr. Worzle accomplished the remarkable feat of becoming a superior stone thrower as a result of being exposed to an unusual challenge. In order to attract the troop to their camp area where it could be more easily observed, Jane Goodall and her husband Hugo van Lawick made a practice of putting out bananas. But this unnatural concentration of food also attracted baboons that lived in the vicinity, leading to abrasive confrontations. The baboons quickly learned which chimps, mostly females and juveniles, they could intimidate by rushing at them. But they never could dislodge Mr. Worzle, who stood his ground, picking up whatever was handy and throwing it at them. Sometimes it was leaves. Once, to the delight of the baboons, it was a bunch of bananas. But slowly Mr. Worzle

82

Positive feedback, a key mechanism in human evolution, is the reinforcement one development gives another: e.g., upright walking (top row) did not itself cause good throwing (second row) but enabled throwing to improve—which aided better walking. The six evolutionary developments traced horizontally (arrows) all interacted like this, though it is not possible to link any step vertically from one row to another.

learned that rocks were best, and as time went on he depended more and more on rocks and started using bigger and bigger ones.

One of the most extraordinary events that took place at the Gombe Stream was the elevation of a chimp named Mike from a very low position among the adult males in the troop to one of dominance. Mike managed this by the remarkable intellectual feat of finding a new—and very frightening—weapon with which to intimidate other males that up till then had been in the habit of kicking and shoving him around with impunity.

Mike's great discovery was that Jane Goodall kept kerosene cans around her tent. He would come into the tent, collect two or three of these four-gallon cans, then deliberately launch an unexpected charge at a group of high-ranking males that were quietly grooming each other nearby.

Charging behaviour in chimpanzees results from the build-up of a high level of excitement or anger, which is released by a violent rush, often upright, accompanied by hooting, arm waving, branch throwing and other aggressive behaviour. Ordinarily a dominant male will ignore the charge of a subordinate by simply letting the other animal run by. But Mike's charges were far too dramatic to be ignored. He would throw the cans ahead of him, kicking two or three of them along as he ran. This made an appalling racket, and the other males would scamper off in terror, bothered as much by the noise as by the size of the cans that were bouncing towards them. This left Mike in command of the field, eyes glaring, panting with fury, hair erect. He had, in effect, "shot up the town". His six-guns were still smoking. One by one, some chimps would come back and make var-

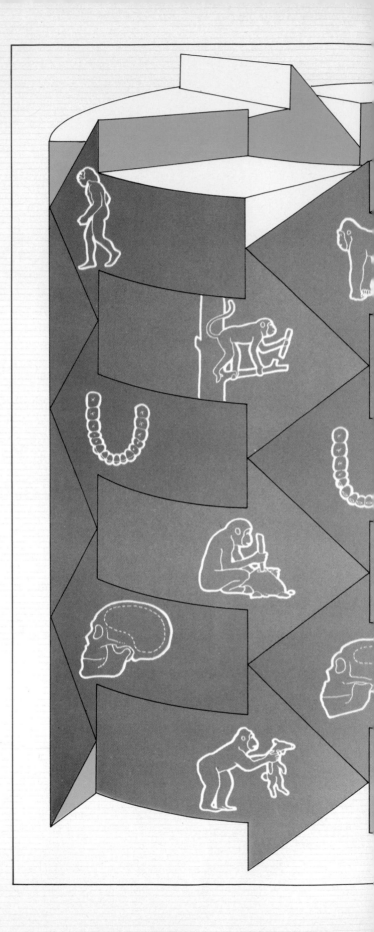

LOCOMOTION

THROWING

TEETH

TOOL USE

BRAIN SIZE

HUNTING

ious gestures of chimp appeasement, touch him cautiously, groom him nervously, creep up tentatively with heads down—do all the things that he used to have to do to them.

After he had repeated his sensational can-banging act a good many times, the Van Lawick-Goodalls concluded that this activity was too dangerous to permit its continuation. They hid the cans. Nevertheless, Mike had made his point. He was now so powerful and respected that he was able, a short time later, to face down the previously dominant chimp, Goliath, by a more conventional display of stamping and branch shaking. From that point on, Mike's position at the apex of the hierarchy was solidified and remained so for several years. What impressed Jane Goodall most, and what impresses me in thinking about it, was her observation that Mike made his charges deliberately. Instead of simply falling willy-nilly into the kind of emotional frenzy that normally spurs a chimp to make a charge, he chose his time quite cold-bloodedly. He prudently collected his cans first, then proceeded to work himself up.

Here, again, the condition was an unnatural one: large tin cans do not ordinarily occur in forests. But Mike's response, like that of Mr. Worzle, the quick-learning stone thrower, indicates the plasticity, the ability to improvise, that exists in an intelligent and physically adept animal when confronted by a new situation or given an opportunity to deal with an old one in a new way.

Given this capacity for tool use and innovative thinking among chimps, and presumably among pre-chimps, accepting that the changeless placidity of the forest environment in which chimps have lived for millions of years has not stimulated their evolution as fast as it might have been stimulated in another environment—given all this, what are the conditions, or where is the environment, in which a similarly endowed animal might have evolved faster?

Many anthropologists think it was simply the ground—open ground at the edge of forests and extending out into grassy parkland and savanna. Here a new environment with new food resources, new opportunities and new dangers could be expected to be explored in a new way by an animal with a new potential. Here our hominid ancestor is believed to have evolved the special qualities that would lead to manhood. But even finding theories to account for that evolution is not easy. If the baboon failed to become human through being a quadruped monkey and thus was unable to develop a tendency towards tool use and bipedalism; and if the gorilla and chimpanzee failed through being comfortably established in their own forest niches and therefore were under no evolutionary stimulus to move out on the savanna to assume an erect way of life, how do we account for the success of the "third ape", as Charles Darwin long ago called our ancestor? To deal with that persuasively, some kind of argument must be found that places hominid ancestors on the ground, at the edge of forests or in the open parkland, and then another argument found for the interaction of that environment with hominid traits to produce a man.

A way of dealing with both these arguments at once is by examining the matter of positive feedback: the reinforcing effect that proto-man's special attributes are supposed to have had in stimulating each other to further and faster mutual development. Positive feedback is a widely known phenomenon. Its effects are obvious in the build-up of unusually large waves in the ocean under the right conditions, or in the progressively greater vibrations that are sometimes built up in engines—where the waves themselves help create bigger waves and the vibrations greater vibrations. There is abundant reason to apply positive feedback to evolutionary events as well. However, there is a problem involved here. It is revealed when the elements that make up a feedback model are broken into questions and answers:

Question: So early hominids used tools, did they?

Answer: We assume that they did. Like chimpanzees, they had the potential, and they brought it with them from the forest.

Question: But what stimulated its development?

Answer: Out there in the open they needed tools to defend themselves.

Question: And why was that?

Answer: Because they had small canine teeth.

Question: And why did they have small canines?

Answer: Because they no longer needed big ones. They were becoming erect, which gave them greater opportunities to use weapons. With weapons they were better able to defend themselves; big canines for defence were no longer necessary.

This is a classic feedback model. Once it is started, it is not hard to see how each element in it will push along the others, including that all-important by-product: development of the brain. The only trouble is that the argument goes in a circle. Canines don't get small because you need tools and a biped stance to protect yourself against having small canines!

This circularity in reasoning has been pointed out by the British anthropologist Clifford Jolly, who notes that the more nearly perfect a feedback model is, the more nearly impossible it is to get started. If every-

thing depends so neatly on everything else, he observes, nothing will happen.

In thinking about this dilemma Jolly tried to find an element that did not depend on the others, something that got its initial thrust from some outside influence. Like so many other anthropologists, he was struck by the remarkable differences between the teeth of the earliest hominids and those of the other apes. Those small canines and incisors—together with their abnormally large molars—had to be explained in some way.

Since teeth and jaw structure obviously are connected with eating habits, Jolly found it logical to seek out a shift in emphasis in feeding that could account for the dental peculiarities of the ancestral hominid—a movement away from reliance principally on fruit to reliance on something else. Noting that fully evolved modern Homo sapiens is still dependent on cereal grains—the seeds of grasses—for his diet, he began to speculate that at some time in the past the ancestral hominid might have begun to eat a good many seeds.

Jolly's examination of this hypothesis is complicated, and some of the dental evidence he adduces is beyond the scope of this book, but its main points fall together in enough ways to make up an intriguing case for his idea.

To begin with, there is the niche itself, the open country—a good deal of it, in fact—and an increased seasonal variability in the weather. Alternating rainy and dry seasons are important if a piece of grassland is to maintain itself. In the tropics, the chief deterrent to the encroachment of the implacable forest is water—too much of it standing in sheets in flat spots here and there at certain seasons, and too little the rest of the time. Both conditions prevent the growth of trees. Grass fires started by lightning from occasional thunderstorms in seasonally dry country also destroy struggling young trees. Still another tree-inhibitor is the nibbling of grazing animals that populate grassland as fast as it is created.

At any rate, open country, with its grains, contains a large untapped source of food for an ape that has enough manual dexterity to sustain itself by rapidly picking, scooping or stripping something as small as seeds. Baboons can do this today; so can some chimpanzees that have adapted themselves to a seasonal open-country life, going out during dry periods when the yield of forest food is low. There is no reason why a hominid ancestor could not have done it.

What is needed to become an efficient seed eater? Very large molars, for one thing, to grind up large numbers of small, hard objects, together with plenty of enamel on the faces of the teeth to withstand the grinding. Also a proper hinging of the jaw to develop power for easy crushing, together with enough flexibility in that suspension to permit the side-to-side motion that is necessary for grinding.

But all the flexibility in the world at the back of the mouth will do no good if rotary movement is limited by the interlocking of large canine teeth in front. If the reader will help himself to a small mouthful of sunflower or poppy seeds and proceed to chew them up, he will notice two things. First, as he swivels his jaw around to grind up the seeds, his front teeth will move as much as or more than his back ones. If he tries to check this front-tooth movement, as a set of oversize canines would tend to check it, he will find the side-to-side movement of his back teeth limited. Second, he will find that the highly arched roof of

his mouth, together with a thick and flexible tongue, acts as an efficient mechanism for constantly pushing the mouthful of seeds back under the molars for further grinding until the food is in a condition to be swallowed. As noted, a combination of extremely large molars, small canines in both sexes, rather small incisors and an arched palate is characteristic of Australopithecines but not of apes.

Here, Jolly suggests, lies the clue to the peculiar dental evolution of the earliest hominids, and also the initial thrust to get the feedback process started. If there is a wealth of small food objects like seeds to be eaten in a new environment, and a selective advantage in the evolution of smaller male canines in order to exploit this diet more efficiently, smaller canines will result.

"But," an alert sceptic will say, "what about baboons? Don't baboons have very large canine teeth for protection on the ground? If they became seed eaters, why didn't they lose those teeth too?"

The answer to this question goes back to the fundamental difference between hominids and monkeys. One has that vital erect inheritance and potential for tool use that the other lacks. If a hominid can develop a talent for protecting himself with weapons, or at least intimidating potential attackers, he does not need those big canines at all. Baboons do need them—and they still have them.

Or, *some* baboons do. At one time there was a baboon species, *Simopithecus*, that apparently spent all its time on the ground eating the roots, blades and seeds of grasses. The animal was common in Africa for about four million years, but became extinct a couple of hundred thousand years ago, presumably as a result of competition from man. The interesting thing about Simopithecus is that in addition to having very large molars the male also had uncharacteristically small canines for a baboon.

Speculating about these peculiarities of Simopithecus, Jolly is able to reconstruct a picture that shows its teeth gradually becoming less like those of other baboons during its long residence on the ground and as a result of its diet there—just as hominid teeth were beginning to become different from those of their ancestors in a similar environment, and as a result of a presumably similar diet.

Jolly has thought long and deeply about Simopithecus, and he has discovered things about it that in his opinion further buttress his argument that early man was also an eater of seeds. He has made an elaborate physical comparison between an extinct ground-dwelling hominid (Australopithecus) and a closely related living relative (chimpanzee) on the one hand, and an extinct ground-dwelling baboon (Simopithecus) and living baboons on the other. The point in the comparison is that the differences between Australopithecus and chimpanzee are very often the same as the differences between Simopithecus and other baboons. In other words, extinct hominid and extinct baboon are alike in the ways that they differ from their closest relative. Looking at this curious situation in another way: if hominids have anything at all in common with baboons, why do they have so much more in common with Simopithecus than they do with other baboons?

Why indeed? Jolly's comparisons, while not conclusive in themselves, do focus on a most interesting parallel, and certainly make an arguable case for the appearance of the "third ape" on the ground and out in the open at an early date; probably not biped to begin with, but with a knuckle-walking inheritance and a talent for tool and weapon use that not only make possible the modification of molars and canine teeth to adapt to a diet emphasizing grain, but also encourage further tool use, manual dexterity and bipedalism—all of which combine to stimulate the further development of the brain. This—at last—produces an erect ape where we want him: on the savanna, and with the kind of teeth that Ramapithecus and Australopithecus fossils say he should have.

Unfortunately, as in everything else in palaeoanthropology, not everybody agrees with Jolly. Sherwood Washburn, for one, does not accept the seed-eating thesis. He thinks diet will not explain how Australopithecus evolved his peculiar dentition: large molars and small canines. In Washburn's view these features must be traced to specific activities—to the gradual increase of tool and weapon use and the development of hunting.

As I interpret the argument, Jolly would agree with Washburn on the importance of tool use and hunting —but mainly to explain how human teeth developed from Australopithecine teeth. Jolly is looking further back; he is interested in finding how Australopithecine teeth developed from ape teeth.

So where do we come out? Seed eating or tool use? Perhaps with a mixture of the two, the former the more important at an early stage, and the latter more important later on. Washburn is certainly persuasive in suggesting that tool use—at some point—became the critical factor in an increasingly swift shaping of man and also in his emphasis on the increasingly important rôle that hunting would play in hominid life. Both of these are important matters that need—and will get—more extended discussion.

Chapter Five: Social Life of the Man-Apes

He that hath no fools, knaves nor beggars in his family was begot by a flash of lightning.—Dr. Thomas Fuller

An erect ape! With hominid teeth!

At the start, nothing more—only a great deal of dispute about the timing, the cause and even the degree of erectness. But in addition to erectness and the way our ancestors achieved it, there are other vitally important characteristics of "human-ness" that must be traced to the unique introduction of a forest ape to the open grassland.

One of the most difficult problems in discussing the evolutionary advance of a creature as complex as proto-man in such a setting is that it involves many inter-related developments and functions, each of which depends on the others and each of which affects the others. Assuming for a moment that a change in diet—away from fruit and towards roots, seeds and meat—may have been the catalyst that set things going, the puzzle still resembles a many-sided closed box. Where does one break into it?

A good place to do so is on the side that might be marked "social organization", by examining the kinds of groups that those first ground-exploring apes probably moved in. But is such a thing—at this late date—possible to know?

Not to *know*, but to make some shrewd guesses at. Since early hominids were most closely related to chimpanzee ancestors, and since they shared the sa-

A male chimpanzee in the Gombe Stream Reserve in Tanzania plays with an infant brother. The ease with which the little one stands on two legs to play reveals the chimps' potential for walking erect—and reflects their close relationship to man. The stance of the larger animal is typical of a knuckle-walking ape, its weight resting easily on curled-under fingers.

vanna environment with baboon ancestors, it may be useful to look for clues in the social organizations of living descendants of both these animals. Although both have differences from each other—important ones—they have interesting things in common. And of these, certainly the most interesting is that their society is highly organized.

Far from being a structureless rabble of rushing, squalling, totally capricious animals, chimpanzee or baboon society is remarkably stable, orderly, usually serene and quiet, with order maintained through a complex interrelationship of five main factors. The first is the mother-infant bond. The second is the animal's age, which regulates its progress from one rôle in the troop to another as it gets older. The third is kinship: the continuing relationship of an animal to a brother or a sister or a mother. The fourth is the adult male-female relationship. The fifth is dominance: how high each animal ranks socially in the group.

Thinking about these five factors, one quickly sees that they are still important regulators of human society. Thus, for a very long time, and in three different species, the game has remained essentially the same; only the scoring and the uniforms have changed. For man, for chimpanzee and for baboon, the problem of life is still largely the problem of getting along with one's fellows in a group.

Berkeley's Sherwood Washburn and Stanford's David Hamburg, looking at primate behaviour from the point of view of an anthropologist and a psychiatrist respectively, recognized the importance of group life when they wrote: "The group is the locus of knowledge and experience far exceeding that of individual members. It is in the group that experience is pooled and the generations linked. The adaptive function of

An infant baboon learns about life from an older juvenile that grabs its tail (left), wrestles it gently to the ground (centre) and pretends to bit

The juvenile lets the infant get in a few mock bites in return. This interplay helps forge friendships that are useful in adult life. Gentleness of

ht). The lesson for the infant: it can get knocked about without being hurt.

is ensured by the ever-watchful presence of the infant's mother (right).

prolonged biological youth is that it gives the animal time to learn. During this period while the animal is learning from other members of the group, it is protected by them. Slow development in isolation would simply mean disaster for the individual and extinction for the species."

"Prolonged biological youth." Think about that for a moment. It takes a male baboon six years to grow up, a chimpanzee anywhere from 10 to 15 years. This slow development is necessary if one of the higher primates is to learn all the things it must to fit itself into the complex society into which it is born. Comparisons that are so often made with ants and bees are relevant only up to a point: these insects do live in highly organized societies, but they learn practically nothing. They do not have to; their responses are genetically programmed for them, their behaviour rigidly controlled. But in a fluid society, where education takes the place of programming, where an individual must deal with a multiplicity of daily choices and varied personal interchanges, a long period of youthful learning is an absolute necessity.

This learning period looks suspiciously like play, and it is. For a chimpanzee, childhood play is the equivalent of going to school. It watches its mother look for food, and looks for food itself. It watches her make nests, and makes little nests of its own—not to sleep in, just for the fun of it. Later, during a long adolescence, it picks up from its peers the physical skills it will need as an adult as well as the more intricate psychological skills of learning how to get along with others—not only how to interpret their moods but also how to express its own moods so that they will be understood. Any chimpanzee that cannot learn to communicate properly with its fellows almost certainly will not live to grow up.

All this time the learner is finding its own place among its peers, first in wholly aimless play, later in more significant activity that will help determine its rank several years later as an adult. In sum, there are two sources of learning—two sets of relationships —that make up primate society. One of these is the family relationship (mother-infant-sibling). The other is the larger relationship of an individual animal to all the other members of the troop.

One of the most striking aspects of a baboon troop is the phenomenon of male dominance. In many of the baboon troops that have been studied so far there is a number one individual to which the other males habitually defer. (Often two—sometimes as many as three or more—animals will team up to hold a top position that neither could hold alone.) Below the top male the other adult males usually arrange themselves in descending order of authority. Although there is always a certain amount of minor jostling for position, and even some prolonged and bitter struggles for places at the top—which is to be expected, since that is where the strongest, ablest and most determined individuals naturally gravitate— nevertheless, a dominance hierarchy, once established, tends to be a remarkably stable thing. The ranking animals move confidently through their days, the others deferring to them as a matter of course in confrontations over food, females, selection of sleeping sites, grooming and being groomed, and so forth.

As a matter of fact, it is the behaviour of subordinates in the troop that ensures its day-to-day stability, rather than the constant ferocity of their superiors. A low-ranking animal is subservient. It knows its place, just as a clerk in an insurance com-

pany knows his place vis-à-vis the chairman of the board. An insurance company would be hopelessly chaotic if the clerks were constantly challenging the board chairman, or if the latter felt compelled every few minutes to rush out into the hall, thump his chest and bellow: "I am the boss!"

The situation is not as clear with chimpanzees as it is with baboons. Among the former, dominance is a relaxed matter; the animals are tolerant of one another and exact status is often blurred. Among the latter it is more rigidly enforced. The difference is believed to stem largely from differences in the living conditions of the two species. A baboon troop moves on the ground, where there is strong selective pressure to produce big aggressive males to protect the troop against predators—or at least to gang up in a show of defence until the youngsters and females can make it to the trees. As a result, there is a marked sexual dimorphism in baboons; i.e., the males are noticeably different from the females. They are much larger (often twice as large), much stronger, have far bigger canine teeth and jaws, are a great deal more combative, are less tolerant of lapses in behaviour or threats to their status. They are also very jealous sexually of their "own" females when the latter come on heat. These traits tend to create an authoritarian society in which a mere stare, much less a show of teeth, by a dominant male will remind a subordinate of the realities of its status.

The authoritarian flavour of baboon society long impressed many observers. Here, obviously, in a power hierarchy with its subtle threat signals ("Look out, I'm very dangerous") and its equally important appeasement responses ("I know it, I mean no harm"), was the glue that held society together and prevented a group of potentially dangerous and extremely highly-strung animals from tearing each other to pieces. The puzzle was why there were not more fights, particularly in large troops of several score animals where the steady penetration of the adult male group by husky post-adolescents, the waxing ambitions of already mature ones and the waning powers of older leaders should have made a continuously stable hierarchy from top to bottom highly unlikely. Finally, why didn't every youngster have to start at the bottom and work its way up?

Deeper study by two baboon experts, Irven DeVore and K. R. L. Hall, gave answers to some of these questions. Better knowledge of blood relationships within the troop began to reveal that its stability did not depend so much on the males as it did on certain high-ranking females. True, those females drew some of their status initially from their association with dominant males, but they also constituted a kind of continuing aristocracy of their own based on family ties: mother-daughter and sister-sister. Once established, this aristocracy tended to perpetuate itself. Those privileged—and usually related—females tended to stick together at the centre of the troop, which was the preferred spot because it was the safest from predators. There they groomed each other sociably in an intimate élite coterie, bringing up their infants in an atmosphere of comfort and security that was denied low-ranking females. The latter were forced to hang about at the edge of the group, apprehensively alert to the possibility of a bite or slap if they did not move aside for a higher-ranking animal. Too timid to try to force their way into the established matriarchy at the centre, they passed their timidity and generally low self-esteem on to their babies.

By contrast, the babies reared by the dominant mothers grew up with a far greater chance of achieving dominance themselves. They sucked in confidence and assurance with their mothers' milk. Their friends from infancy were other well-born youngsters. They went to the right schools, as it were, made the right contacts, had all the right opportunities for successful baboon careers.

Chimpanzee social life is even more complex than that of baboons, because rôles are not so stereotyped and there is more opportunity for individual expression. Since they live in the forest and are largely relieved of the threat of ground predation, chimps need not be as combative nor as cohesive as baboons, and they are not. Nor, for the same reason, do they exhibit the degree of sexual dimorphism that baboons do. While they do have dominant hierarchies, these are not as rigidly enforced. Chimp society is more open, more innovative, more relaxed, more tolerant. It is free of sexual jealousy, marked instead by a casual promiscuity. When a female is in oestrus and anxious to mate, any interested males in the troop will line up and amiably wait their turn—which should come soon, since the sex act itself takes only a few seconds and is performed casually, sometimes while offhandedly munching a banana, and even with curious youngsters climbing up and tugging at the performing male.

In an animal of the chimpanzee's intelligence, with a wider range of responses than are possible in a baboon, and a corresponding potential for greater complexity in the relationships between individuals, there seems sometimes to lurk the beginning of an awareness of others. Chimps are probably 99 per cent self-centred. And yet they are caught often enough

in a lordly sharing of food (usually if they have had enough themselves) or in some similar act, to raise the possibility that they may be dimly responding to the needs of their companions. A dominant male that has chastised or frightened another will reach out to touch it, just to reassure it. Family ties tend to be strong and long-lasting. One reason for this is that chimps mature very slowly, and the infant is closely associated with its mother far beyond infancy, and often with any older brothers or sisters it may have. It is tempting to find in these gossamer-thin hints the beginnings of family formation, love, altruism and other attributes that we have learned to regard as exclusively human. But jumping to these conclusions is dangerous. All we can say is that an animal *like* this began to live more and more out on the open ground, and began having to live in an increasingly different way in order to survive there.

We take the family unit—father, mother, children —for granted, naturally enough, for it has been central to human evolution for so long that we tend to forget that there was a time when it may not have existed in its present form. If, in searching for hominid family beginnings, we use as a likely model the chimpanzee family, with its relaxed affections, its prolonged mother-infant and sibling-sibling ties, then we must account somehow for the introduction of a father into the unit. For in chimpanzee society there

Baboons communicate with one another by using a variety of signals. On the page opposite, a dominant male yawns—not an expression of boredom but a threat underscored by a show of teeth and light-coloured eyelids. On this page the male baboon at right acknowledges its subordinate status by "presenting" its rump to a dominant male, which then reassures the other by a casual pat on the rump.

are no fathers. Where friendly promiscuity is the rule, fathers cannot be identified, nor are they needed. There is plenty of food in the forest (no need for a paternal provider), and almost no predators (no need for a paternal protector). Where, then, does the father come from?

Does he appear as a result of extended sexual attraction between males and females? It is scarcely necessary to point out that humans differ from the casually promiscuous chimp in that they routinely form permanent male-female bonds, and enjoy a mutual attraction that is continuous and does not depend on the female's monthly cycle at all. But how and when humans achieved this is an utter mystery. One suggestion is that among animals that are as easy-going as chimps and enjoy physical contact—grooming, touching, stroking, often just sitting or lying close together—any change in the environment or the social structure of the troop that has the effect of throwing a male and a female together over an extended period of time might cause the slow development of a continuous sexual interest between them. Another idea is that the periods of the female's sexual receptivity could gradually become longer and longer under the stimulation of such conditions of extended male-female intimacy until the periods themselves overlapped, making her continuously receptive.

But that period of extended intimacy between a male and a female—where does it originate? It does not exist among chimpanzees, where males prefer the company of males. It is absent in the East African savanna baboon. However, it is present in some other baboon societies, notably among the Gelada and the Hamadryas species. Both of these animals live in open country that is not only generally drier but also has greater seasonal swings in climate and in the availability of food than the country familiar to the savanna baboon, a forest-edge animal that finds itself living in an environment of relative food abundance the year round.

Here environmental difference coincides with differences in social organization, a coincidence observed by the noted British scientist John H. Crook of the University of Bristol. He has studied the social organizations of many animals, including certain African weaver birds and antelopes as well as baboons. What has struck him is the apparent uniformity with which *all* these otherwise entirely different creatures react to similar environmental change. So striking were his findings that he has used them as the basis for a hypothesis: *Under similar environmental conditions social animals will tend to develop similar social organization.*

Crook's conclusions about social organization are based on complex and subtle evidence that is beyond the scope of this volume. However, a look at three different African baboon societies should make clear the point of his argument.

Baboons are extremely adaptable animals. They have not become unduly specialized physically, and are thus able to fit themselves to a wide variety of living conditions. This is to be expected in any animal of generalized physique and considerable intellectual attributes. Man himself is the best example of this. Thanks to his brain—and its by-product, culture—he can live near the North Pole with a body that is essentially the same as the body of a man living on the Equator. With no "culture" to rely on, baboons must fall back on changes in their social organization in order to adapt to different living conditions.

Of all African baboons, the East African savanna species lives in the easiest surroundings—close to the forest to which it can flee and in whose trees it can sleep, and in a benign climate where there is year-round food abundance. In this setting the troop is the all-important social unit. Family relationships, except for mother-infant ties, are dim. The nearest thing to a father is a dominant male that, as a matter of course, exercises the *droit du seigneur* over some or all of the females in the troop.

By contrast, Geladas are confined to mountain slopes in Ethiopia. There the climate is harsher, seasonal change greater, and food availability more chancy. As a result male-female relationships change. The troop breaks down during the day into a bunch of wide-foraging separate units, each containing a single adult male with one or more females and assorted young. The logic in this arrangement is obvious: during times of food shortage adequate food for females and young is far more important for the survival of the species as a whole than is food for extra males. So long as there is one strong male to protect the females and impregnate them during oestrus, other males can be regarded as so much surplus baggage—useful only for replacement or for forming new families of their own with young females. With this social structure the relationship between a particular male and female is far more durable than it is among savanna baboons, and in that respect it resembles the human nuclear-family unit more closely than does the savanna-baboon troop—or even the chimpanzee troop. Significantly, when the wet season rolls around in the arid Ethiopian hills and food starts to become more abundant up there, the one-male groups break down and begin to coalesce

into larger, more conventional multi-male troops.

Hamadryas society is different again. This baboon lives in even drier country than the Gelada, in rocky sections of Ethiopia and the near-desert of Somalia. In this environment one-male groups are the rule the year round. The male-female relationship is more close-knit than among Geladas. Each Hamadryas male is continuously jealous of its harem, requiring its females to stay very close to it at all times. When it moves, they move—or get bitten. This particular behavioural trait is so ingrained that a female Hamadryas, when threatened by her male, will always run towards him, not away.

Now, hominids are not baboons and are not even very closely related to them. Nevertheless, Crook's thesis is provocative. If social organization is shaped by environment—and he would have it understood that the shaping does not take place directly but over a long period of time, initiated by and reinforced by selection—then it can be argued that an ape, moving out of the forest to live in seasonally dry country, could have modified its social organization to conform to the requirements of that environment. Furthermore, it could have done so in the way that other primates are known to have done. These modifications obviously would have varied from place to place. Just as baboons have differing life styles, surely hominids did too, depending on where they lived and on how severe the seasonal food- and water-getting problems were. Where such problems were at their worst, extra male hominids may be considered to have been just as expendable as surplus male baboons, and one-male family units may have resulted.

This supplies a father-figure for our small-toothed, open-country, erect ape—a figure that was missing

in the easy-going chimpanzee-like society that hominids most probably had before they left the forest. And it was certainly not the peculiarly human characteristic of round-the-clock sexual receptivity that introduced the father to the "home" (although, later in his evolution, such receptivity may have become one of the forces that encouraged him to keep coming back to it). On the contrary, at the stage of which we are speaking—as we try to identify the beginnings of a human family structure in a creature that is not yet a human—the determining factor may well have been environmental. If so, the long-term association of a male with certain females was an economic one. Under certain conditions it was more efficient for survival if one-male family groups evolved.

This search for a father has been roundabout, and the evidence for the explanation offered here is entirely circumstantial. Nevertheless the search for some kind of an acceptable argument is necessary. We have heads of families now, and have had for a long time—they didn't just suddenly jump onstage; therefore they must be accounted for. Also, the father—or more properly, the male family head—plays a key rôle in so many interrelated developments that are peculiarly human that the emergence of man simply cannot be conceived without his presence at some point far back in time.

The question really becomes: how far back? Crook's baboon model would seem to make the male-

Reassurance and close contact are important to chimpanzees. On the opposite page the female (left) shows anxiety with a "fear face", its teeth bared, and gets a reassuring gesture from a male. On this page several chimps groom one another (top), a favourite pastime and a source of comfort to them. In the bottom picture a group huddles companionably in a nest.

female bond a very old one, since it is based on the fundamental issue of environment, and therefore could logically be expected to begin to manifest itself soon after the first exposure of hominids to a ground life where seasonal food shortage was any kind of a problem. That could take us back to the earliest of the Australopithecines or perhaps even to pre-Australopithecines, but here guesswork takes over.

Those who disagree with Crook say that it is not necessary to look that far back, or as far afield as baboons, to explain family formation. They prefer to keep their eye on that closer relative, the chimpanzee, and attribute the beginnings of family formation to meat eating and food sharing—both traits that chimps display in a feeble form. That, says Sherwood Washburn, a strong advocate of this view, was the influence that led to the development of permanent male-female bonds. Since the eating of meat leads to an improvement in hunting techniques, and since the practice of hunting begins to get involved with more effective tools and weapons and with better bipedalism, the Washburn model implies a somewhat later date for "family" formation.

Be that as it may, everyone is agreed that the rôle of the male as family man in hominid evolution is an important one. It has had its effect on the evolution of different (and appropriate) male and female rôles in daily living; on the development of teaching elaborate (and appropriate) new skills to the young, on the development of the concept of a home base, also on the matters of hunting and food sharing. All, obviously, are interconnected. Together they make for a complex feedback system.

Rôles suggest evolutionary shaping to fit those rôles. For example, men are usually bigger and stronger than women now, and they surely have been for millions of years. These attributes are predictable for protectors and hunters; the relationship between rôle and physique is simple and direct. But men can also run faster than women, and here the reason is neither simple nor direct. If speed afoot were merely a matter of size and strength, then the largest and strongest man would be the fastest runner. Since this is demonstrably not so, then there must be another reason why a lithe woman should not be as fleet as her bulkier mate.

She is not for two reasons. First, she does not have to be; the rôle that she will increasingly assume as mother, homemaker and gatherer of food does not require her to run fast. Second, she cannot be if she is going to be the mother of larger-brained children. The best-designed pelvis for the bearing of such children is not the best-designed for running—or even for the most efficient walking.

To achieve good bipedalism, there had to be evolutionary changes in the shape and proportions of the proto-hominid's foot, leg and pelvic bones, also in the muscles of leg and buttock. A chimpanzee can walk quite comfortably on its hind legs, but not for any length of time. It can also run surprisingly fast. But for efficient running and walking, as men know it, the chimpanzee simply is not built properly. Its legs are too short, its feet are not the right shape, its big toes stick out to the side like thumbs instead of forwards to give spring to the stride, its three principal buttock muscles are rather small and poorly placed. What muscles it can put to use for walking are attached to its bones in such a way that insufficient leverage is provided to achieve a vigorous stride. Also, the chimp proceeds in a kind of waddle or rolling gait, because its legs are so far apart that it must shift its body weight at each step to position it over the leg that is on the ground.

A more efficient way of walking erect is to have the legs straighter and closer together, as they are in man. The male reader should be able to make his thighs, his knees and ankles all touch each other when standing with his feet together. This is made possible by a marked change in the shape and proportions of the leg and foot bones, and particularly the ancestral pelvis. There has been a twisting and flattening out of the two large pelvic flanges, which not only helps set the trunk more vertically on the legs, but also provides better fastening and far better leverage for the three sets of gluteal—or buttock—muscles that are used in walking. Men have gradually achieved all these improvements. Chimpanzees and gorillas have not; hence their straddle-legged and inefficient shuffle.

Changes in man's pelvis, important as they are for walking, have not made the open space in its centre any bigger. In order to achieve that, the whole pelvic structure itself would have to get bigger—which would defeat the evolutionary process of developing a compact, efficient pelvis for bipedal walking. It is here that the hominid female is presented with a hard choice. Her pelvis cannot be compact enough for the most efficient walking and running without its being too small to permit the passage through it of the head of her baby when it is born. In fact, as hominids have become more man-like and larger-brained, the pelvic opening should really have become even larger than it is. That it did not is the cause of the difficulties in childbirth that the modern woman experiences. Her pelvis is a compromise.

When hominids entered the savanna, we can be sure that they, too, brought with them clearly defined differences in the rôles played by males and females. Since hominids were more intelligent than baboons, their families were even more closely tied down by infant care. Not only did their infants mature more slowly, but they were becoming more helpless at birth. This is because one of the solutions to the small-pelvis, large-brain problem is to eject the infant into the world at an earlier stage in its development—before its head gets too big—and sentence the mother to an even longer period of infant care.

With influences like these at work, and evolutionary pressure for still bigger brains being steadily applied by increased tool use, it is logical to assume an intensification of the differences in the rôles played by males and females, particularly when the females became more and more tied down by their infants and more dependent on the males with whom they were beginning to associate longer. Longer association encourages mutual support. New behavioural twists become possible, one of them being the slow beginnings of food sharing.

Baboons do not share food, but chimpanzees sometimes do. Hominids may also have the traces of that trait when they left the forest. On the savanna, with males becoming increasingly good walkers and extending the range of their activities accordingly, the chances of their finding small animals to kill, and later beginning deliberately to hunt them down, obviously grew. And the incentive to share food must have grown accordingly. One cannot eat an entire baby antelope on the spot. What one might well do is share it with other hunters then and there, and carry what is left back to the slow-moving, infant-

Revealing behaviour strongly reminiscent of a human, a chimpanzee mother at the Gombe Stream Reserve cuddles a week-old infant.

encumbered female with which one is permanently associated. Since hunting involves chasing prey, sometimes for considerable distances, some group members must be left behind, most probably females and infants. "Behind" should at best be a place where those less mobile members of the group are reasonably safe, and at worst, a place that the hunters can find again—in short, the beginnings of a home base.

Hunters are not always successful. As often as not, they return empty-handed. As a result, in nearly all hunter-gatherer societies they supply only a fraction of the group's food. Therefore it is the females' responsibility to see to it that there is a dependable supply of fruit, seeds, nuts and other vegetable materials. In addition to evening out the ups and downs in the food supply created by those undependable hunters, such things as nuts and grain, unlike quickly rotting meat, keep a long time and may be rationed out in small quantities as needed.

In some such way, we can assume, the sharing of food got its start—along with the development of appropriate male and female rôles for procuring it. These are rôles that would endure for some millions of years. They are, in fact, still commonly found in hunter-gatherer societies today.

How far this division of labour had progressed by Australopithecine times is debatable. Certainly its progress was uneven. For a female to be a useful gatherer of small food objects she must have something to put them in, something she can carry about with her. This implies the use of baskets or gourds or containers made of large leaves or pieces of animal skin. There is no evidence whatsoever that Australopithecines used any of these, but the lack of evidence does not mean they didn't use them, for all such materials

are perishable. Basket use had to start some time. The picture that has been put together of Homo erectus and his culture makes it pretty clear that he was capable of making containers of some sort.

Could Australopithecus have done this? All that can be said is that some time prior to one million years ago female hominids had already begun the collecting and keeping of more food than they could eat at the moment themselves. Did this start two million years ago with Habilis or three million years ago with earlier gracile types? Most experts consider that the Australopithecine brain was too small to engage in making and using containers. And yet, the existence of rather sophisticated toolmaking techniques, recently demonstrated by Mary Leakey, should make anthropologists extremely cautious about letting fly with firm pronouncements on what early hominids could or could not do. As with Jane Goodall's chimpanzees, the more we learn about Australopithecines the more capable they seem to have been.

What has all this speculation produced?

It has produced an image of a social hominid with both a group structure and, possibly, the beginnings of a family, its society intricately organized, probably on dominance lines originally. This hominid has moved from the forest to open country, there exploiting an ever-widening range of foods, including seeds. In short, it eats anything and everything it encounters. It has brought with it the ape's potential for bipedalism, tool use, and meat eating. On the savanna there are real advantages in developing these traits, and it exploits them all. It is an efficient erect walker by Australopithecine times, and may have been erect before that. Distinct male and female rôles, of protector-hunter and homemaker-gatherer respectively, are beginning to assert themselves. These rôles —and here we find ourselves once again in the circle of positive feedback—are not only necessary for protection and sharing in order to take care of larger-brained, slower-developing infants and their mothers, but are also made possible by the larger brains that this new way of life, with its reliance on sharing, bipedalism and tool use, encourages.

In this connection, the mere ability to walk easily becomes significant. Washburn has pointed out that many primates never move more than a few miles from the spot where they are born, and thus are sentenced to a necessarily parochial view of the world. Even a baboon, keen-sighted though it may be, and smitten by who-knows-what baboon visions as it gazes out from its tree-top roost upon a vast world of many mysteries and unknown places, is never galvanized by those visions. It habitually keeps to a range of 10 or 15 square miles. Efforts by K. R. L. Hall to drive a baboon troop were easy so long as it remained inside the small area that it considered home. But trying to push the troop beyond its home boundaries always failed; the animals would double back into country whose every tree and rock they knew and where they felt safe. In choosing safety, of course, they gave away a chance to learn.

For hominids, whose potential ranges began to grow as their physical ability to patrol these ranges increased, the opportunities for new sights and experiences grew correspondingly. So did the selective pressure to evolve larger brains capable of storing more and more information about that larger world. With increased movement came the stimulus to carry things for greater distances, itself a further stimulus to biped walking—and to further exploration.

There is absolutely no way at present of calculating the range of an Australopithecine band, and it must surely have varied from place to place and from good year to bad. Nevertheless, it was almost certainly measured in the scores, and may well have been in the hundreds, of square miles. With that expanded range went all the advantages that could accrue to a mobile group living in it, able to move from one end of the area to the other when necessary, to escape local disasters of drought or flood, to take advantage of local bounty as it developed seasonally, and, most important, to be able to remember where, when and how to exploit that large domain. Simply by increasing the options available a large-brained hominid that can walk respectable distances increases its chances for survival.

That covers most of the arguments about the earliest hominids except size—their own size and the size of the bands they moved in.

By mid-1971 the total number of known Australopithecine fossils, including some unpublished finds by Clark Howell, Yves Coppens and Richard Leakey, had grown to 1,427 (*page 46*). However, the vast majority of these finds are teeth and bits of jaw. The tell-tale long bones of arm and leg, which are the only way of making accurate size estimates of Australopithecines, are still extremely scarce and fragmentary. Nevertheless, they permit some rough guesses, which come out something like this:

The Boisei male, which appears to have become larger and more robust during a period of several million years, ranged up to five and a half feet tall, and weighed as much as 200 pounds—a big animal. Boisei females seem to have run 15 to 25 per cent smaller and may have weighed only half as much.

Africanus, if we include all the gracile types, from the smallest South African specimens to the larger and later Habilis ones, ran in a range of four and a half to five feet in size and 80 to 100 pounds in weight. Again, females were somewhat smaller.

Robustus, so far found only in South Africa, falls somewhere between the Boisei and Africanus populations. Males were about five feet tall and weighed up to 150 pounds. Females were about four and a half feet tall and weighed about 80 pounds.

Estimates of the number of Australopithecines in a band range from a dozen or so individuals up to 50. These estimates are based on the known sizes of chimpanzee, gorilla and baboon troops, also on the practical size limits that would be put on (and, incidentally, are still put on) hunter-gatherers by the day-to-day difficulty a larger group would have in finding food and water for all its members.

As to how long Australopithecines lived, that too is a subject for speculation. A chimpanzee has a life expectancy of about 25 years in the wild, although it has the physiological potential to live as long as 60 years. Australopithecines, of similar size and close lineage, could be expected to have the same potential. But as is the case with chimps, this potential was apparently severely reduced under actual living conditions. The palaeoanthropologist Alan Mann has been examining the rate of tooth development among immature specimens of some South African gracile fossils and the rate of tooth wear among adults of that same population. His studies indicate that none of them got past 40 and that only about one in seven lived as long as 30 years. Their mean life span, he estimates, was about 20 years—which says a good deal about the hazards of Australopithecine life.

Movement to the savanna may have led to the evolution of human beings, but it would be millions of years before this would have any pay-off in easier living or greater longevity.

Chapter Six: Weapons and Tools

Five young lions rise to stalk approaching antelopes—a hunting technique that man's predecessors, the Australopithecines, may have used.

*It is useless for the sheep to pass resolutions
in favour of vegetarianism while the wolf remains of a
different opinion.*—Dean Inge

During the years that they spent at the Gombe Stream the Van Lawick-Goodalls watched some close friends among the chimpanzee troop grow old and die, some strapping adolescents begin working their way up the male hierarchy, some babies born. And while they were observing all this, they had a child of their own. This infant, nicknamed Grublin, was taken back to Tanzania, lived at the Gombe camp as a baby and still spends much of the year there with his parents. But fond as she was of the chimps, and knowing that many of them were fond of her, Jane Goodall was still extremely careful about leaving Grublin by himself when he was small. She knew that the possibility always existed that one of the chimpanzees might casually kill and eat Grublin.

She had discovered that her friends were occasional meat eaters and also occasional hunters. Her first suspicion of this came when she saw a chimp in a tree working over something pinkish red, and two other chimps sitting close by with their hands out in a begging gesture. They were rewarded with a piece of that something—later identified by her as the carcass of a young bush pig. The chimpanzees were eating meat, a trait utterly unsuspected by her. As time passed she caught them eating meat on several occasions, and even observed them hunting, a practice confirmed by Japanese observers who began studying chimpanzees in Tanzania in 1961.

A hunting chimp, Jane Goodall reports, is unmistakable. There is something a little out of the ordinary about its behaviour, something purposeful, tense, inward that she can recognize and that other chimps also recognize and respond to. Sometimes they just watch intently. Sometimes they move to adjacent trees to cut off the escape of the quarry, a young baboon or small arboreal monkey. On several occasions the quarry was a young baboon whose screams usually brought adult baboons rushing to its defence. In the ensuing hullabaloo the youngster more often than not escaped. But she saw chimps eating infant baboons often enough to realize that there was a small but fairly steady toll taken of them through the year.

Chimpanzees are excited by meat and clearly very fond of it. They chew it long and reflectively, usually with a mouthful of leaves added. Wads of this mixture are occasionally given to other begging chimps. Sometimes the carcass itself is shared by the owner, who will tear off pieces and hand them out. The curious thing about meat eating and meat sharing is that regular dominance patterns do not apply. A ranking chimp that would not hesitate to assert itself over a lowly one for possession of fruit will respect it in the matter of possession of meat. Apparently there is something about having killed an animal oneself that gives one the right to it.

The revelation that chimps hunt and eat meat—*and share it*, albeit often reluctantly—has enormous implications in explaining the development of hunting and sharing among hominids. It now becomes possible to speculate that these traits were brought to the savanna from the forest. We no longer have to puzzle over how a propensity for meat eating got started in a creature with a fruit-eating ancestry; it was there, as it is in a great many animals. All it needed was encouragement in a new environment.

Although the development of agriculture and the dramatic rise of civilizations in the last five to ten thousand years tend to obscure it, the fact is that our ancestors have almost certainly lived by hunting and gathering for more than a million years, perhaps for two or three million years. They have been marvellously successful at it, and many of our physical characteristics and some of our most deeply ingrained emotional traits stem from our long careers as hunters. Out of the last three million years of their evolution as bipedal hominids, our ancestors probably have spent 99 per cent of their time as hunters. The modern non-hunting way of life, which we complacently regard as *the human* way, becomes a single gasp in a whole day of breathing.

But before turning to weapons and tools as evidence of what kind of hunting was done, let us consider briefly the practice itself: hunting as a way of life, how and to what extent it may have developed in the early years of hominid existence.

One good way of approaching this subject is to look at it through the eyes of George Schaller, who asks us not to concentrate so hard on primates and primate behaviour for a moment, but to look at other species that roam and hunt the African savanna. Schaller has written: "Since social systems are strongly influenced by ecological conditions, it seemed [to me] that it might be more productive to compare hominids with animals which are ecologically but not necessarily phylogenetically similar, such as social carnivores."

Social systems . . . influenced by ecological conditions? That makes sense. We have already heard it from John Crook in a slightly different context.

But social carnivores? Is he talking about *lions?*

Yes, he is. Acknowledge that the earliest hominids did carry a potential for upright walking, tool using and meat eating with them from the forest. But try to understand that when they became ground-dwellers living in groups out in the open, the slow development there of those traits might be best explained by looking at the activities of other meat eaters that also lived in groups out in the open.

Of the largest African carnivores—lion, leopard, cheetah, spotted hyena and wild dog—all but the leopard and cheetah are social animals that have developed two vitally important traits: they hunt in groups and they share their food.

Co-operative hunting has many advantages. Schaller lists no fewer than five that give a group of hunters a big edge over an individual working alone. First, the group is more successful, day in and day out, in killing. Two or more spotted hyenas, working together, kill an animal at the end of a hunt more than three times as often as a single hyena does. Second, a group can kill larger animals than one hunter can. The most dramatic example of this is probably the wild dog, which, working as a member of a pack, can bring down zebras that weigh up to 500 pounds, though the dogs themselves weigh only about 40 pounds each. Third, *all* the food caught by a group will probably be eaten on the spot, with none wasted. This is often not possible for a single animal, which eats as much as it can and then must wait until it is hungry again. By that time other predators may have found the carcass and pre-empted it. That is why the leopard, a solitary hunter, is put to the trouble of hauling its kills up into trees where hyenas, jackals and dogs cannot reach them. Fourth is what Schaller calls division of labour. Here he cites the example of a wild dog that will remain behind to guard pups in the den while others are off hunting for food that they will bring back in their bellies and regurgitate for the pups —and also for the adult guard. Conversely, one lion will lie next to a partly eaten kill, guarding it until the other members of the pride arrive to finish it off. Finally, there is what might be termed *force majeure*. There is a power ranking on the savanna based on size, with the lion at the top, the leopard next, followed by the hyena and wild dog. But numbers can overbalance size. A photograph on pages 114 and 115 shows a frantic lioness unable to defend the carcass of a giraffe against a dozen hungry hyenas.

I would add a sixth advantage that Schaller does not include in his list, but that he is well aware of and, indeed, may consider covered under his heading "greater success". That is the wider range of hunting methods that is open to a group than to a single animal. Among these is a kind of relay-race effect achieved by wild dogs. One or two adults will start a chase by running right after the prey, keeping constant pressure on it. As it flees, it is apt to run in a wide circle, which enables trailing dogs to trot slowly along, watching the action, then save steps by cutting across the arc of pursuit and close in for the kill. Another example: lions are very good at driving prey towards partners lying in ambush. They are also adept at surrounding it, so that whichever way it bolts there may be a lion in a position to take a swat at it. A group hunting together can sometimes manoeuvre the quarry into a cul-de-sac: out onto a promontory, into a swamp or a river, into a gully from which it cannot escape. Hominids, through their history as hunters, have used all these techniques.

Another very important aspect of social carnivore

behaviour is food sharing. While lions fight and snarl —and sometimes even kill each other—over their food (a trait that suggests incomplete social evolution; *i.e.*, they may have learned to co-operate in the field but not yet at the table), hyenas and wild dogs are much better behaved. Dogs, in fact, are extraordinarily scrupulous in this regard. The pups in the pack, being slower runners, are apt to be the last to arrive at a kill. The adults usually content themselves with a few bites, then move aside to let the youngsters have their fill before settling down for their own meal. This sometimes means that they go hungry and must hunt again, but good care of pups has great survival value for a species in which mortality among adults seems to run high. When wild dog pups are too small to follow the pack and must stay in their dens, adults return, and in response to nibbles and pokes at the corners of their mouths, will regurgitate food for the little ones to eat. A lame dog that could not keep up with the hunt practised the same method of begging for regurgitated food and was kept alive by the generosity of its pack mates.

Co-operation and sharing, then, confer strong benefits on social carnivores. For a hominid venturing out into the open these same benefits could accrue. The more widely he ventures, the greater will be his chance of stumbling over small prey animals such as hares, fledgling birds and the newborn young of larger herbivores. Not only will he be increasingly stimulated to chase and kill these food objects but, even more important, he will be encouraged to be on the lookout for them and to think more and more about how and where to find them. His ambitions will gradually grow as he realizes that crippled or old individuals of even larger species are within his capabilities as a killer. But the bigger the game, the greater the need for co-operation. And with proportionally larger amounts of meat on his hands as a result of successful co-operation, the greater the opportunity and the incentive to share it.

Here, again, positive feedback will operate. The more successful a particular kind of behaviour is in an animal intelligent enough to remember and to have some choice in how it behaves, the more apt it will be to continue to try what has worked before. Each animal caught strengthens the urge to look harder for more animals. This phenomenon was observed by Jane Goodall as she watched the ebb and flow of embryonic hunting activity among the Gombe chimpanzees. A chance catch of a young baboon would stimulate a good deal of hopeful hunting energy. But because of their poor success as hunters, their pell-mell and misguided enthusiasm, and because of the large number of alternative food sources open to them, the chimps would quickly be discouraged by successive failures. Hunting would cease to be the "in" thing until the next lucky kill rekindled interest.

Stimulated by more frequent success in the open, and perhaps also by a greater need to hunt and scavenge because of a less stable food supply out there, hominids could have organized something that was unimportant to the survival of chimps into something that was important to them.

Sharing, just as much as co-operative hunting, would have improved the survival chances of a hominid. It is painful to watch a sick or injured baboon trying to keep up with the troop. Other baboons do not feed or look after it in any way; being largely seed, grass, fruit and root eaters, they have to spend most of the day feeding themselves. Therefore the

Porcupine

Oryx

Sivatherium

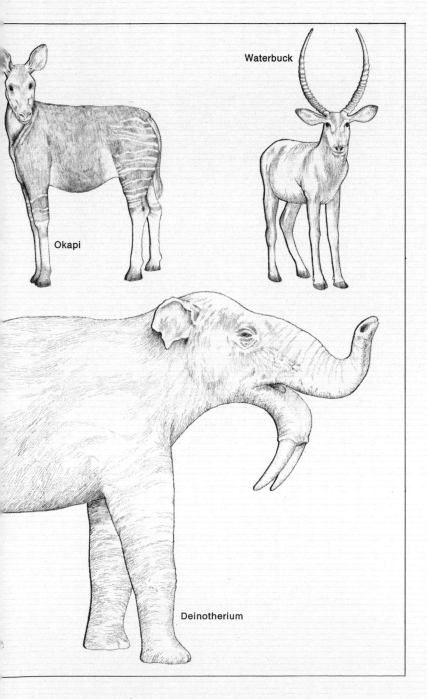

Okapi

Waterbuck

Deinotherium

sick individual must manage on its own, and even though the other animals may be moving slowly through the day, the disabled one still may not be able to find the energy to forage because it is devoting all its strength merely to keeping up. In such a situation it will get weaker, find it harder to keep up, get still weaker, and so on.

The bringing back of food to a place from which a disabled individual does not have to stir for a critical few days might mean the difference between life and death—particularly for a hominid whose maturation and learning processes are very slow, and whose experiments with a meat diet may expose it to the ravages of unfamiliar internal parasites that might infect it. A baboon with a broken leg or a prostrating case of dysentery is almost certainly a dead baboon. A similarly stricken hominid might survive.

So the peculiar attributes of a hominid interact with the co-operative hunting and food sharing of a social carnivore. The result, in its earlier stages, could produce something like an Australopithecine, a hunter that goes about its business in a new way—on two legs and with weapons—its wits constantly being sharpened by that new way until it eventually becomes a very efficient hunter indeed.

But that "eventually" was certainly slow coming, so slow that it was probably a long time before hominids became efficient enough even to be recognized by other animals as dangerous. As hominid skills in-

Bones found in hominid dwelling sites at Olduvai tell what those early hunters ate. Four of these prey species still exist in Africa; oryx, porcupine, okapi and waterbuck. Two are extinct; sivatherium, a primitive giraffe with a short, thick neck and curved horns: and deinotherium, a kind of elephant with a shorter trunk than the modern species, as well as a strange lower jaw with down-curving tusks.

Like the hunting dogs in the action sequence opposite, early hominids are believed to have been able to cut a weak animal from a herd—as in this drawing of a gazelle capture. The hominids were unable to run fast themselves but were probably able to drive such quarry into an ambush or a cul-de-sac, or simply to wear it down by chasing it relentlessly.

creased, the secret undoubtedly became harder to keep. By late Australopithecine times, two or more million years ago, it is almost certain that early man was a sufficiently skilful hunter to make all but the largest herbivores afraid of him. The big carnivores, lion and leopard, preyed on him. Packs of hyenas may have "mobbed" him, for in large groups they are very aggressive. But the hominid hunter was probably aggressive too. It is likely that he competed with hyenas and wild dogs, fighting with them over their kills, his kills and over the found carcasses of large dead animals. In those confrontations, numbers and aggressiveness undoubtedly decided the issue.

Still, hunting started on a very modest scale, limited to the chance finding of small animals. Equally important, probably in the earliest stages, and possibly for a long time thereafter, was scavenging—the finding of game already dead, either from natural causes or killed by other animals capable of being intimidated and driven away. This might be called the opportunistic side of meat eating, and here, once again, hominids resemble social carnivores, which are past masters at opportunism. Although a lion, encountering hyenas on a fresh kill, will drive them off, the hyenas are capable of gathering reinforcements and will drive away the lion. Two or three lions will turn the tables again.

Another fascinating aspect of the social life of carnivores is the diversity—and low degree—of dominance among them. If animals are to co-operate during hunting, aggression between individuals must somehow be discharged or suppressed. But in a linear hierarchy this is very difficult. It is hard to imagine a group of status-conscious baboons swallowing their mutual animosities and fears to the point

Text continued on page 116

ild dogs co-operate to stampede zebras, hoping to flush a laggard.

They succeed in cutting out one zebra that is weaker than the others.

s the rest of the zebra herd runs off, the pack closes on the quarry.

A well-organized hunt ends as the dogs begin to devour their victim.

Hunting in parties, hominids were much more formidable than they would have been alone. They are believed to have been effective scavengers, using bluster—shouting and waving clubs—to drive carnivores away from kills that they had just made. Here a hunting party intimidates a lioness.

The effectiveness of numbers is clearly demonstrated in this excepti

...tograph of a lioness that has killed a giraffe but is unable to defend it against the mobbing tactics of a menacing crowd of hungry hyenas.

Food sharing is another adaptive advantage savanna-dwelling hominids (above) are thought to have had in common with social carnivores like lions and wild dogs (opposite). Bringing a small carcass back to females and infants, as wild dogs do, or sharing a large kill, as lions do, increases hunting efficiency enormously; it ensures that all the meat so laboriously caught will be eaten and lessens emphasis on storing a kill.

where all can run shoulder to shoulder in a co-operative hunting venture. But carnivores do it. Among lions males are dominant over females, but only because they are stronger. The females do not accept this lying down, and often fight back in disputes over food. There is no hierarchy among females themselves, who do most of the hunting. Among hyenas, females are dominant over males, but again the sexes are without hierarchies. Wild dogs have a permissive and amiable society in which dominance varies from pack to pack, but is nowhere strong. Where it exists it seems to be largely an expression of relationships between individual animals.

These patterns are in marked contrast to the dominance hierarchies of many primates. Whether, in looking for a hominid model, one selects the genetically close chimp or the ecologically close baboon, one does find dominance as a structuring influence in social life. At one time hominid society *must* have been dominance-orientated (it still is in many respects). Nevertheless, for it to have become a successful hunting society, it must have changed. David Pilbeam suggests that aggressive behaviour among male hominids began to disappear as a result of the emergence of pair-bonding between males and females. He also believes that the beginnings of language—a more sophisticated level of communication that could convey something more than mere feelings—would lead to greater trust, greater understanding and greater co-operation between individuals. Language development, he says, "would have made possible for the first time in primates the reward and reinforcement of nonaggressive behaviour patterns. Dominance ceased to be its own reward."

ild dogs share food by bringing it to pups and regurgitating.

ree lionesses and a male share a kill brought down in a stream.   The solitary leopard keeps its kill for itself by dragging it up into a tree.

On this point there is some dispute. Pilbeam's remark carries the implication that language was developed very early, possibly as early as Australopithecine times. Other experts disagree. While they concede that language may be a useful moderator of aggressive behaviour—you curse somebody or complain instead of hitting him with a club—they do not concede that it was necessary in stimulating nonaggressive behaviour. That, they maintain, resulted from family formation, pair-bonding, long mother-infant relationships and food sharing—all this long before language came along. In fact, they insist, Australopithecines were too small-brained to have been able to talk. They contend that speech—other than a vocabulary of sounds to convey feelings such as alarm, anger, pain or pleasure—was beyond the capacity of hominids before the emergence of Homo erectus a little over a million years ago.

Furthermore, early hominids may not have needed to talk. The real value of language—in addition to the enormous stimulation it gives to brain development—is that it permits the conveying of subtle things that are beyond the power of grunts and gestures to communicate. The latter are themselves surprisingly subtle, and do permit a remarkably high degree of communication among such animals as chimpanzees. But while we may assume that an Australopithecine knew more than a chimp, and thus had a need to communicate more, how much more is hard to say. Like everything else, language origins were gradual and slow in coming, and it would be impossible (if we could re-create the actual events—which, of course, we can't) to distinguish between what was a remarkably eloquent and informative series of sounds and what was a piece of true speech. We simply do not know, and never will, how or when language began.

Since this whole matter is highly tentative, it may be more useful to cut across it with Schaller's sensible observation that speech is not necessary during hunting. Carnivores do not communicate while so engaged; indeed, some of them hunt at night, using stalking techniques that require silence and render visual communication difficult. Wild dogs hunt during the day, also in silence, except for an occasional yelp to help keep the pack together. Other signals are unnecessary, since the chase unfolds in full view.

Because apes and monkeys are diurnal animals, Schaller—like virtually everybody else—assumes that the early hominids were too, and that they did all their hunting and scavenging in daylight. This is overwhelmingly logical. For one thing, the night is dangerous: a small hominid strolling after dark would have been all too likely to run foul of sabre-toothed cats, lions, leopards or hyenas, all of which hunted at night. For another, hominids were extremely keen-sighted in daylight. Assuming that they were also biped, they were tall enough when standing up to see considerable distances, and also mobile enough to cover a good deal of ground. This suggests that they kept a sharp eye out on what was going on around them and that much of their activity was directed towards scavenging. Too slow afoot to run down healthy large animals by themselves, they probably relied on hyenas and wild dogs to do that for them, then ran up in a noisy group and drove the hunters away. They probably could have run down weak or aged herd animals by themselves.

Having made his trenchant points about social carnivores, Schaller prudently backs off with the observation that there are many styles of hunting among them and that at present there is no way of telling which, if any, were exactly followed by hominids—or even if different hominids did different things at different times. Nevertheless, the analogies are there, and they are extremely provocative—so much so that Schaller decided to turn himself into an Australopithecine for a few days to find out more about what they may or may not have been able to do.

Selecting the Serengeti Plain and one of its rivers as an environment whose climate and large herds of grazing animals are still very close to the presumed conditions of a couple of million years ago, Schaller and fellow scientist Gordon Lowther conducted two experiments as hunting-scavenging hominids.

The first experiment was conducted on the open prairie where the two men walked, a hundred yards apart, for a total distance of about 100 miles over a period of several days. Their main target was baby gazelles, which, in the first week of life, do not run but crouch down in the grass. They found eight and could easily have caught them all. An excellent haul, but there was a catch to it: five of them were spotted within a space of a few minutes at a place where pregnant females had gathered to deliver their fawns. Since birth among most of the plains herbivores is concentrated into very short periods—with the survival advantage of simply overwhelming predators by the sudden production of more new-born animals than they can possibly eat—Schaller concluded that catching gazelles was a poor long-term proposition: marvellous for a few days, poor the rest of the time.

However, he and Lowther stumbled over some other things on that same walk: a hare they could have caught, a couple of adult gazelle carcasses partly eaten, a cheetah making a kill a mile away—which they could have pre-empted. Conscientiously adding up all the bits and pieces, including some scraps of brain from one kill gnawed almost clean, they came up with about 75 pounds of meat.

Their second experiment was conducted in a woodland strip edging the Serengeti's river Mbalageti, and lasted a week. There they had better luck. They hung around the riverbank where the herd came to drink—and ran into competition from 60 or 70 lions that were also hanging around. They found four lion kills, but these had all been picked so clean by the lions that there was nothing left but some brains and the marrow in the larger bones. This latter, as tool-using hominids, they could have recovered by smashing the bones open with rocks.

They found a partly eaten buffalo that had died of disease, and could have recovered about 500 pounds of meat from it—a big strike. In addition they found an 80-pound zebra foal, abandoned and sick, also an oddly acting young giraffe that they discovered was blind when they succeeded in running up and catching it by the tail. It weighed about 300 pounds. But the game supply, then as now, was fickle. It rose and fell seasonally, according to drought, disease and migration. To cope with this, hominids must have been under some selective pressure to become increasingly artful hunters, to learn how to stalk, ambush and kill healthy animals when there were no old and sick ones to be had. Then there were always seeds, nuts, roots and fruit for them to fall back on—just as hunter-gatherers do today. But to make a direct comparison between the Australopithecine hunting-gathering life and that of certain modern hunter-gatherers like the Kalahari bushmen is wrong, according to Schaller.

The bushmen have been pushed into a semi-desert area where there is almost no game at all. The bulk of their food has to be vegetable.

Small and unintelligent though they were, early hominids probably got more meat than modern bushmen. But whether they did or not is not the important matter. What counted in the long run was the activity itself. The challenges of hunting are in themselves stimulative to the brain. One of the strongest influences in the intellectual evolution of man, as Sherwood Washburn keeps emphasizing, was undoubtedly his activities as a hunter—although Washburn thinks it is a mistake to look to the social carnivores for hunting models. He maintains that one need look no further than the chimp—with its hunting proclivities—to account for a hunting tradition in early hominid behaviour. That was germ enough, according to Washburn. It changed our ancestor by enlarging his horizons and his mental capacity. He gradually learned to hunt better, to think and plan better, and to use and make better tools.

For hominids, which lack great speed, great strength and great canine teeth, tools make the hunter. Their origin in human evolution is hidden forever in the misty processes of trial and error. The best we can do is remind ourselves that there was a time when our ancestors could do less with tools than chimpanzees can do now, and that they must have worked their way up through a similar—but not necessarily identical—limited capacity to shape something for a purpose: a grass stem for poking into a termite mound, a chewed-up mouthful of leaves to serve as a sponge, a stick or branch as something to be brandished in an effort to intimidate, a rock to throw.

For a bunch of not-too-large apes, standing erect will make one seem more formidable because it makes one appear to be larger. The brandishing of sticks or branches will enhance that effect, and may have been enough, on occasion, to swing the balance in a clash with hyenas over possession of a kill. The earliest use of implements by man's ancestor, as a ground-adapted scavenger-hunter, may have received its strongest impetus from its value in threat displays against competing species.

The found implement—whether stick or stone —was obviously the only implement for an immensely long time, picked up and then thrown away when its immediate use was over. But there must have come a stage at which Australopithecines (or their ancestors) began to recognize more and more clearly the usefulness of certain objects, and, as a result, tended to hang onto them longer—to the point where they may have begun carrying them around much of the time. This, as Washburn has suggested, may have been the great incentive to bipedalism. The more you want or need to carry things, the more you will walk on your hind legs. The more you walk on your hind legs, the freer you are to carry things.

Stones are easy to find and easy to throw, perhaps only for purposes of intimidation at the start, but eventually with a growing realization that they can actually damage, and even kill, if they are thrown hard and accurately. Hitting with a club is probably easier yet. The great abundance of wood, and the fact that it is softer and easier to work than stone (that is, until the craft of stone-tool making is understood and mastered), suggests that the earliest hominids used wood a great deal, and also the long bones of some of the larger animals. But the great triumph of our ancestor as a beginning creator of culture is in the legacy he has left us of worked stone. Most of that stone, it should be pointed out, survives as implements, not as weapons.

The magnet that drew Louis and Mary Leakey back to Olduvai Gorge year after year was the existence there of large numbers of extremely primitive stone implements. Mary Leakey has made the study of those objects her special province and has published a splendid monograph on the stone culture at Olduvai. This covers material taken from Olduvai's lowest strata—known as Beds I and II—and a time period that extends from not quite two million years ago to about one million years ago.

What Mary Leakey has managed to reconstruct of the lives of the creatures who lived there at such a tremendous remove in time is stupefying. That scattering of mute stone tools is so old and so cryptic that it might have been expected to keep its secrets forever. But she has made the stones speak. She has found exactly where hominids lived. She has learned a great deal about what they did. She has even found what appears to have been a shelter of some sort that they built. She knows what they ate, and where they ate it. Her findings represent more than 40 years of uninterrupted work: the collecting, the sifting, the recognizing, the exact plotting in position, the describing and the interpreting of hundreds of thousands of bits of material—some stone, some bone; some large, some extremely small—no one of which, taken alone, would mean much. But when they are all analysed and fitted together like a gigantic three-dimensional jigsaw puzzle, patterns begin to emerge that speak across the gulf of time. These patterns turn a group of small, two-dimensional, cardboard, scarce-

ly believable anthropology-book creatures into real —I cannot call them animals, I cannot call them people—into real living beings.

The first thing Mary Leakey does is classify the stone culture itself. In general, she finds that there are two stone-working traditions in Olduvai. One is the Oldowan, the older and more primitive of the two, producing mainly what was for a long time called pebble tools, but what she prefers to call choppers. The word "pebble" suggests something quite small, and her term is an improvement, for many of the choppers at Olduvai are of hen's-egg size or larger, some of them three or four inches across.

An Oldowan chopper is about the most basic implement that one can possibly imagine. It is, typically, a "cobble", a stone that has been worn round by water action—like so many that are found in the beds of streams or along rocky seashores. It is made of some close-grained, hard, smooth-textured material like quartz, flint or chert. Many of those at Olduvai are made of hardened lava that has flowed out of the volcanoes in the region.

A roundish cobble, then, oval or pear-shaped, small enough to fit comfortably in the hand—that is the material for an Oldowan chopper. What an early toolmaker had to do to turn it into a tool was simply to smash one end down hard on a nearby boulder, or balance it on the boulder and give it a good whack with another rock. A large chip would fly off. Another whack would knock off a second chip next to the first, leaving a jagged edge on one end of the tool. With luck this edge would be sharp enough to cut up meat, to saw or mash one's way through joints and gristle, to scrape hides, to sharpen the end of a stick. There were large choppers and small ones. There

CHOPPER, A CUTTING TOOL EDGED ON ONE SIDE

were also the flakes that had been knocked off in making the choppers. These were sharp too, and were also used to cut and scrape.

The Oldowan industry is found in Bed I. It extends up into Bed II, improving somewhat as it goes. But Bed II also contains traces of a more advanced industry, the Acheulean. Its characteristic implement is the biface, a tool whose cutting edge has been flaked more carefully on both sides to make it straighter and sharper than the primitive Oldowan chopper. Also, the Acheulean tool is often worked or trimmed all over to give it the desired size, shape and heft. The result is the hand axe, the basic implement of the lower Stone Age.

The astonishing thing about the Olduvai tool kit is not that it evolved—that is to be expected—but that it was so elaborate. Mary Leakey has identified 18 different kinds of objects in Beds I and II. Among them, in addition to choppers and bifaces, are round stone balls, scrapers, burins (chisels), awls, anvils and hammerstones. In addition there is a huge amount of so-called debitage, or waste: the small flakes and chips

that would naturally accumulate in a spot where implements were being made over a long period of time. Finally, there are manuports: stones that show no workmanship at all, but were carried by hand to a site, evidence for that being that stones of like material do not occur in the locality. A handsome white ostrich egg of a rock picked up on the beach at Martha's Vineyard and used as a doorstop in a home in Ohio is a manuport.

Choppers, picks, cleavers, awls, anvils, spheroids? What on earth are we talking about here? Can this really be the detritus of a culture left by a two-million-year-old hominid that most anthropologists believe was too small-brained even to talk?

Indeed so. That is the fantastic picture that Mary Leakey's work reveals. Just as the intricacies and subtleties of chimpanzee or baboon society turn out to be much more complex than anyone realized a generation ago, so does the culture of the early biped hominids. The reason the Leakeys concluded that the creature responsible for turning out this elaborate tool kit was a man—and therefore should be called *Homo habilis*—is the elaborateness of the culture Mary Leakey unearthed, not brain size. She does not care how big a hominid's brain was, but she cares strongly about what he could do with that brain. If he could make tools—not merely use them, make them to a regular pattern—then he was a man.

Incidentally Mary Leakey does not adhere to the view held by many anthropologists that Habilis is an Australopithecine descendant. She believes he goes back on his own line and that the gracile types are cousins, not ancestors. This argument is a subtle and difficult one, but it is essentially semantic. It will probably be settled in time, not so much by reshuf-

*PROTO-BIFACE, A CUTTING
TOOL EDGED ON BOTH SIDES*

*PICK, FOR PIERCING
AND DIGGING*

*ANVIL, USED AS A BASE
FOR CHIPPING OTHER TOOLS*

fling the relationships of fossils as by agreeing on a set of names for them that will satisfy everybody.

Perhaps the most mind-boggling of all the things Mary Leakey has turned up at Olduvai is "occupation floors". These are places where hominids actually stayed for extended periods of time, dependent on the local vegetation and the local game. They are, in effect, two-million-year-old homes, and are identifiable by heavy concentrations of fossil material, stone tools and debitage in a small area, and confined to a depth of only a few inches. The earth those hominids squatted on is still there, with the re-

SPHEROID, A TYPE OF HAMMER

HAMMERSTONE, FOR
MAKING OTHER TOOLS

HAND AXE, FOR DIGGING, CHOPPING OR CUTTING

mains of what they made and ate scattered all about.

Gradually, blown dust, encroaching vegetation, rising water and mud covered each of these occupation floors, but gently and without disturbing anything. Thus, the objects that the Leakey's have been so laboriously uncovering and cataloguing have remained exactly where they were dropped by those who dropped them. Elsewhere in Olduvai the artifacts and bones are spread through layers of sand and clay that may be several feet deep. In these places it is clear that river action has moved them about, swirled them together and dumped them from time to time, and that their position relative to one another is not very meaningful. But as one lays bare an occupation floor, one has the sense of going down into a basement to look at the rusting tools, the stacked storm windows, the jam jars on the shelf, the pile of comic books, the lawn mower and broken electric fan—all left just so—and, in looking, to learn something about the life that their owner lived.

What did Homo habilis leave in his basement? For one thing, he left a lot of fish heads and crocodile bones, together with the fossilized rhizomes of papyrus plants, indicating that at one site, at least, he was living next to the water and getting some food from it. Other sites contain the bones of flamingoes, which says that the nearby water was a lake, that it was shallow and slightly alkaline—as many East African lakes are today—since only such conditions produce the tiny water creatures flamingoes eat.

Ten occupation sites have been identified at Olduvai, out of about 70 that contain fossils or tools, scattered along a 12-mile stretch of the gorge. One floor has its cultural debris arranged in a most peculiar manner. There is a dense concentration of chips and flakes from tool manufacture, mixed in with a great number of small smashed-up animal-bone fragments—all of it crowded into a roughly rectangular area some 15 feet wide and 30 feet long. Surrounding this rectangle is a space three or four feet wide where there is hardly any of this cultural junk; the ground is nearly bare. But outside that space, the material becomes relatively abundant again. How can this extraordinary arrangement be explained?

The most obvious explanation is that the densely littered central section was a living site, that it was surrounded by a protective thorn hedge, that the hominids who lived there relaxed safely inside that hedge while they made their tools and ate their food, and that whatever they did not simply drop right there on the floor they tossed out over the hedge.

At another site is a roughly circular formation of stones about 14 feet across. This is as attention-getting as a trumpet blast on a summer night. Not only are there very few other stones on that occupation floor, but what stones there are are widely and haphazardly scattered. By contrast, the circle is a dense concentration of several hundred stones carefully arranged in a ring by somebody—somebody who also took the trouble to make higher piles of

CLEAVER, FOR SKINNING OR CUTTING

stones every two or three feet around the circle.

That this configuration should survive after nearly two million years is staggering. It suggests a shelter, of a kind that is being made today by the Okombambi tribe of South West Africa. They, too, make low rings of stones, with higher piles at intervals to support upright poles or branches, over which skins or grasses are spread to keep out the wind.

Although the predictable debitage of flaked debris is found inside the stone circle, indicating that some activity took place there, there is more evidence of a wider variety of activities having taken place outside. This makes sense. The interior dimensions of this somewhat irregular circle are only about eight feet by ten or twelve feet, which would have made things rather crowded in there if it had been the home of several people. Furthermore, that group had in it some extremely good hunters or scavengers. The surrounding area contains the fossilized remains of giraffes, hippos, many antelopes and the tooth of a deinotherium, an extinct elephant. Those people were eating a lot, and may have found it more convenient to do their eating out in the open rather than in the confines of the shelter.

Whether they actually killed those large animals, whether they chased them into swamps and helped them die, whether they brought home meat from found carcasses, whether they pre-empted the kills of other carnivores, the chronicle of Olduvai does not say. But it makes clear that when an extremely large carcass became available they did cut it up and eat it. There are two sites in Olduvai Gorge known to have been butchering sites. One contains the skeleton of an elephant, the other that of a deinotherium. Since those animals weighed several tons each, it was obviously impossible to move them; the thing to do was to settle down at the carcass and chop and chew away at it until its meat was gone. On the evidence at these butchering sites, that is exactly what happened. At each there is an almost complete skeleton of a huge animal, its bones disarranged as if they had been tugged and hacked apart. And, lying among the bones are the discarded choppers and other stone tools that did the hacking.

The Olduvai hominids were very catholic eaters. Certain sites are rich in antelope bones, some with their skulls cracked open at the precise point on the front where the bone was thinnest. Others are crammed with the shells of large tortoises. One is littered with snail shells. Another contains a giraffe head, but nothing else belonging to that giraffe —clearly it was lugged in to be eaten at home. A site higher up in Bed II reveals an increasing dependence on horses and zebras, which means that the climate had become drier by that time and was encouraging the spread of open grassland. There is also a marked increase of scrapers in Bed II, which suggests the beginning of an effort to work hides and leather.

The clues are many, and enthralling. What do we make of little concentrations here and there of very small bones, most of them broken into tiny pieces? Was any hominid so quixotic as to collect handfuls of skeletal fragments of mice, shrews, small birds and lizards and then carefully place them in piles? It seems wildly unlikely, and Mary Leakey has concluded that these strange little heaps are probably what is left of hominid feces. This means that our ancestors were eating those small animals whole, bones and all, much as a modern man might eat a sardine. The bones were ground up into very small fragments by chewing, then passed through the intestines and deposited right where they were found.

Mrs. Leakey's attention to these minute details is extraordinary. At one site she has collected more than 14,000 bone fragments, so small that all together they weigh only 15 pounds. Her measurement and classification of tools are equally painstaking. She can tell exactly the "mix" of 14 different kinds of tools at any one of the important sites in which she has done extensive work. This kind of study reveals that the chopper was overwhelmingly preferred in all the Bed I Olduvai sites. But, moving up into Bed II, the spheroid becomes the most common implement at most places. What on earth were those round stone balls for? They are too carefully made, and represent too big an investment in time and labour, to have been used simply as throwing missiles—they could have been lost too easily. Mary Leakey believes that they may have been used as bolas. The bola is still seen on the South American pampas. It consists of two or more stones connected by thongs or cord. These are whirled around the head and thrown at a running animal or large bird. Not only is the chance of hitting a target far greater with a whirling bola that may be two or three feet across than with a single thrown rock, but the weapon is also very effective in tangling up an animal's legs. Furthermore, if it misses entirely, it can be found and used again.

With this wealth of fantastic information coming from Olduvai, there can be no question that by two million years ago hominids were living in a remarkably advanced state of culture—at a level that no one in his wildest dreams would have believed possible a few decades ago. Since progress in the early Stone Age was abysmally slow, this means that the begin-

nings of the Oldowan tool industry are far older than Olduvai—how much older no one yet has the slightest idea. But in 1969 word began to trickle back from East Rudolf and Omo that there were tools there too. The first official news of this came in 1970, when Mary Leakey published a paper describing some implements from her son Richard's East Rudolf dig at Koobi Fora. The next year two experts went to Koobi Fora to help Richard Leakey analyse his site. They were Glynn Isaac, a prehistorian from the University of California at Berkeley, and Kay Behrensmeyer, a geology student from Harvard. Their analysis confirmed another stunner: an occupation floor containing animal bones and Oldowan choppers and flakes that is probably three-quarters of a million years older than Olduvai.

What is particularly promising about Koobi Fora, along with several other nearby sites in East Rudolf, is that when the geology and dating of the whole area are worked out, and one place properly linked to another in time, it will then be possible to connect directly the area's exceptionally rich hominid fossil remains with its tool industry finds—and to learn more about Australopithecus as a maker and user of tools 2.6 million years ago.

We know—logic forces us to accept—that toolmaking and hominid evolution go hand in hand. But the earliest steps are still totally unknown. They bring us back to the shadowy beginnings of hominid history, to a time when it might indeed have been virtually impossible to distinguish between a worked tool and a found one, to a time when even something as primitive as Ramapithecus might have been experimenting with stone implements while it pondered the advantages of walking erect.

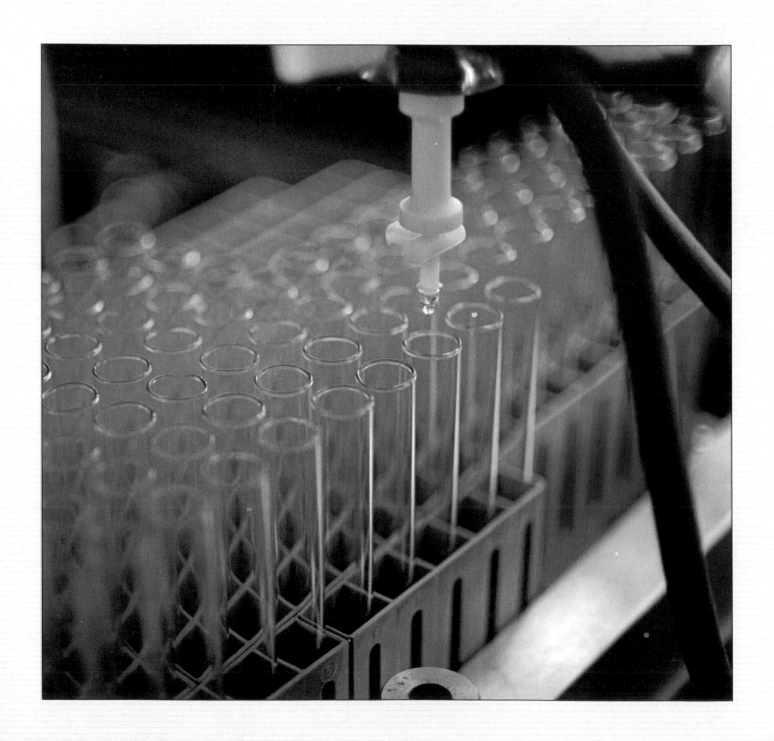

The profoundest truths of the Middle Ages are now laughed at by schoolboys. The profoundest truths of democracy will be laughed at a few centuries hence, even by school teachers.—H. L. Mencken

Sherwood Washburn, when talking about hominid evolution, will often declare: "My prejudice is. . ."—meaning that what he is about to say next is subject to dispute. It is a polite way of warning his listener that he may be about to speculate or grind some special theoretical axe of his own. Having no axe to grind myself, except the basic one that the origin of species by natural and sexual selection is at the root of man's appearance on earth, I have felt free to wander about, picking up scraps of late news and following whatever lines of argument seemed the most reasonable to me. The preceding chapters, then, reflect, more or less, "my prejudice".

Having admitted prejudice, it is only fair to admit further that there are arguments that produce conclusions quite different from several that have been reached in this book, and that some of my prejudices are more shakily based than others.

For example, I have tried to dwell equally upon fossils and primate behaviour. This will not satisfy a pure fossil man, who will argue that geology, bones and artifacts are the only hard evidence of evolution. They are the proof that such-and-such happened, because here they are waiting to be measured, dated

The search for man's origins is not confined to digs in the field. The drop about to fall into a test tube in the Carnegie Institution's microbiology lab in Washington, D.C., contains strands of genetic material from a man and a chimpanzee; analysis reveals similarities between them—and the intimacy of relationships among their prehistoric ancestors.

and compared to other objects. In this view, to waste one's time with engaging but unprovable fantasies based on the behaviour of modern primates, and particularly of animals even less closely related to man, is all very well, but it does not advance the science of palaeoanthropology very far.

In response to this the behaviour-orientated man will say: We are very closely related to some of those animals, and while our behaviour is obviously not identical with theirs, it springs from a common source, and to ignore the many matters that come up from an analysis of animal behaviour is simply narrow-minded. Furthermore, how much can be learned from those static little bits of bone? Palaeontologists are constantly making deductions from dental minutiae and just as constantly finding themselves disagreeing.

To see how far afield a concentration on one discipline to the exclusion of all others may lead, let us consider the matter of the primate family tree: just when who split off from whom.

Recently the Finnish palaeontologist Bjorn Kurten addressed himself to this problem. Looking only at fossil evidence he concluded that men are not descended from apes at all. He based his reasoning on examination of certain small-jawed 30-million-year-old primate fossils. One in particular, known as *Propliopithecus*, has been under periodic scrutiny by palaeoanthropologists as a possible ancestor of man. Concentrating on the jaw and teeth of this animal, Kurten decided that we can draw a direct line of descent from it, down through Ramapithecus to Australopithecus. His argument is simple: Small jaws and teeth, not large ones, are the primitive condition. Man resembles the early, generalized, small-jawed model more closely than do apes or monkeys with their later specialized larger jaws and longer canines. Assuming that Propliopithecus is a hominid ancestor, said Kurten, it is hard to construct an evolutionary scenario that has canines getting bigger (to accommodate the development of apes) and then getting smaller again to account for the later split-off of man. He prefers a small-jawed ancestry all the way, saying, in effect, that monkeys and apes are "descended from men," not the other way around. Kurten's scenario would put the split between hominids and apes some 30 or 40 million years back.

This, of course, would demolish the argument, developed in previous chapters, that a small jaw and highly molarized teeth may be the result of environmental adaptations to living on the ground, adjusting to a diet of small, hard seeds, hunting, sharing and other matters. Considering that the simplest explanation is usually the best, Kurten's argument would seem superior to the one developed here.

Logical as it may be, however, the Kurten scenario, suffers from a shortage of actors. It attempts to string together a line of fossils that exist in such scarcity and in such fragmentary condition that not enough is known about them to speak with any security.

Nevertheless, the fossils—scarce and unsatisfactory though they may be—are there, and Kurten is not alone in trying to make something of them. There are other bone men who interpret them in ways that put the split between hominids and apes at 30 million, 20 million and 15 million years. There are behaviour men who arrive at similarly diverse results, but for different reasons. There are combination bone-behaviour men; they cannot agree either.

Since it is impossible to make strict scientific measurements of either fossils or behaviour because both vary from individual to individual, the arguments will go on. Both fossils and behaviour are, in the last analysis, slippery, elastic yardsticks. Many scientists feel that what is needed to measure evolution more precisely is something as inelastic and dependable as radiometric dating, something that consists of small measurable units—like those decaying atoms of potassium-40 in volcanic ash—units that do not change, that are found in all living things and that can be counted in the laboratory.

It appears that there may be such a way of measuring evolution. What all living creatures have in common—in the form of measurable units—is genes, mysterious substances in the hearts of all cells that determine what those cells will be. Will a fertilized egg become a bumblebee or a buffalo? That question is answered by the egg's genes, which also determine which cells of the growing egg will become buffalo hair, which will become buffalo legs and which will become a small wart on the back of the buffalo's neck.

The rôle of genes in directing cell growth had been suspected since the turn of the century. But *how* they did this job was an utter mystery until 1953, when two future Nobel laureates, James Watson and Francis H. Crick, directed their attention to a nucleic acid in cells called DNA. They succeeded in unravelling the structure of DNA and establishing that this was the agent that told genes what to do.

DNA acts like a punch card giving information to a computer. The virtue of a punch card is that it can be used over and over again in any number of computers and will always give the same result. Another virtue is that an enormous amount of information can be stored on the card by using only one kind of storage unit—small holes in the card. The holes themselves are all alike, but their position on the card can be changed. That is what counts, for every such position change results in a different message from the card to the computer.

DNA acts on much the same principle; it, too, can store an unbelievable amount of instruction by using simple units. Instead of being a card, it consists of two long strands of chemical building blocks strung together like beads on a chain and then wrapped around each other in a spiral pattern. This is the famous "double helix" discovered by Watson and Crick. It, too, has virtually endless potential for variety. Instead of using one kind of unit—a hole—it uses four kinds of chemical building blocks. The different instructions it gives are determined by the arrangement of those blocks—like the holes in a punch card—on the two strands of the helix. Equally important is the way the two strands, as they twist into the double helix, are firmly held in that spiral relationship by chemical bonds.

Since DNA instructs genes, and since genes direct the growth of bee or buffalo, it begins to be clear that if evolution is to occur, it has to start with changes in the arrangement and the linkages of the building blocks on the double helix.

Such changes do occur in DNA. They occur in the form of mutations, and, for simplicity's sake, we might call these changes "units" of evolution. As time passes, changes will slowly begin to accumulate in the DNA of any species. Eventually these changes will become numerous enough and effective enough so that the species will begin to show the effects of them. At that point what we describe as evolution can be said to have taken place.

The longer the time, the greater the number of those evolutionary changes will be. It follows, then, that if a way could be devised to measure the differences in the DNA of two different animals, it would be possible to measure the evolutionary difference between them—simply by counting. In other words, by comparing the DNA of a man to that of a chimp it should be possible to see just how closely related they really are.

This sounds simple in theory. In the laboratory it turned out to be fiendishly difficult, since it involved developing a method of taking apart, looking at, then analysing and putting back together strands of submolecular size—objects so small that they are invisible under the highest magnification of conventional microscopes. Despite the difficulty, molecular biologists like David Kohne of the University of California at San Diego and B. H. Hoyer of Carnegie Institution do it. They take advantage of the fact that DNA comes in two strands. A single strand from the DNA of a man can be unwrapped and matched to a single strand peeled from the DNA of a chimp.

It turns out, according to the laboratory measurements used, that man differs from chimp by only 2.5 per cent, from gorilla only slightly more. But he differs from monkeys by more than 10 per cent. As Washburn, who has been following this work closely, says: "You can construct an accurate family tree from DNA without ever once looking at the animals themselves. In fact, the tree for primates published by Kohne in 1970 would have caused no surprise to Darwin and Huxley in 1870."

Why do it, then? Why go to all that trouble to prove something that science has already shown to be true in other ways and has accepted for a century?

Because it provides evidence of a sort that did not

Studying Molecules to Unravel the Evolutionary Tangle

THE DNA METHOD

Human DNA

Gorilla DNA

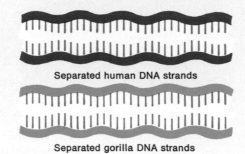
Separated human DNA strands

Separated gorilla DNA strands

Combined human-gorilla DNA

THE PROTEIN SEQUENCING METHOD

Human haemoglobin protein

Differences in haemoglobin protein

Gorilla haemoglobin protein

THE IMMUNOLOGY METHOD

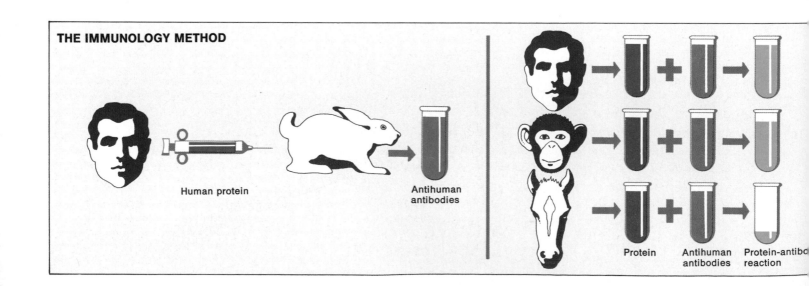

Human protein

Antihuman antibodies

Protein

Antihuman antibodies

Protein-antibody reaction

Scientists, trying to calculate how closely related various species are to one another, have worked out three basic techniques in the laboratory that accomplish this by measuring differences in the species' DNA and protein molecules. The DNA method, diagram on the left, involves genetic material, whose full name is deoxyribonucleic acid, and takes advantage of the lucky circumstance that its molecules consist of two strands made up of simple compounds. The strands are wrapped around each other in a double spiral, or helix, and held in position by strong chemical links between the strands.

It is possible to break the bonds between strands in the laboratory—to unwrap the double helix—and thus disentangle one strand from the other. If this is done with DNA of a man and a gorilla, and a single strand from each is permitted to recombine, all of the chemical links between the two strands will re-establish themselves except at the points where the links are chemically different—indicated in the drawing by two gaps where the links oppose each other. Inasmuch as those differences represent mutations—genetic variations that produce evolutionary change—the matter of how closely man and gorilla are related can be determined by the number of chemical links that fail to re-establish themselves. It is such differences in their DNA that make one creature a man and the other a gorilla.

A second method of determining the evolutionary "distance" between two species is to compare protein molecules such as those in blood. All protein molecules are made up of the same building blocks—20 different amino acids in long chains, connected to one another in different orders. Man, mouse and gorilla are all made up of the same amino acids; it is how the acids are arranged that determines which is which.

Complicated laboratory techniques now make it possible to examine a protein molecule from one end to the other and determine for that protein the exact sequence of the 20 amino acids as they occur over and over again in different arrangements. For example, haemoglobin, the red-blood protein, consists of a string of 287 units of amino acids, whose arrangement has been worked out for a number of animals. The more alike the sequences are, the closer the relationship of the animals, the more different, the more distant the relationship.

In man and chimp the amino acid sequence of haemoglobin is identical. Man and gorilla are very close; their haemoglobin has only two differences. By contrast, the haemoglobin patterns of man and horse differ at about 43 points. In the simplified diagram left, symbols have been used to indicate only six different amino acids instead of the 20 that actually exist in haemoglobin protein. Arrows indicate points at which differences occur.

Protein sequencing, though precise, is extremely arduous because proteins may contain hundreds of the same 20 amino acids in different orders. The immunology approach bypasses the exacting task of identifying one amino acid after another. It relies on an animal's capacity to build up antibodies to defend itself against foreign elements introduced into its bloodstream. Antibodies causing reactions with one creature's proteins will cause similar reactions with proteins of closely related animals, but hardly any reaction in distantly related ones.

If some serum-albumin, a blood protein, is taken from a man and injected into laboratory rabbits, they will produce antibodies to combat this foreign stuff. This procedure is illustrated at far left, with the antibodies coloured orange. Serum containing the antihuman antibodies can now be used to gauge relationships between humans and other animals. Combined with human serum-albumin, this antihuman serum (*top row*, *near left*) will react violently, having been manufactured by the rabbit specifically to combat human serum-albumin. This reaction is represented by the full blue test tube.

Chimpanzee serum-albumin, being only slightly different from a human's, will cause almost the same violent reaction (nearly full blue test tube). But serum-albumin from a horse is quite different from a man's, and will affect the antihuman serum very little.

	MAN	GORILLA	GIBBON	MONKEY
MAN	—	8	14	32
GORILLA	8	—	14	32
GIBBON	14	14	—	32
MONKEY	32	32	32	—

	MAN	CHIMP	OLD WORLD MONKEY	NEW WORLD MONKEY	TARSIER
CARNIVORE	173	173	174	172	155

These tables plot differences in blood serum albumin among primates (left) and between primates and carnivores (above). The fewer the differences—as, for example, between man and ape—the closer the evolutionary relationship.

exist before. Now the units of measurement are the same. No matter where the experiment is conducted, what chimp is compared to what man, the difference will always be constant. Now, at last, there is a standard yardstick of evolutionary separation that is fundamentally unlike those elastic differences in brain size or tooth size—which will always vary slightly from individual to individual. In making DNA comparisons in the laboratory there is *no* variability from one test to another.

Molecular biology does not stop with DNA. It has gone on to use other related laboratory techniques to investigate the evolutionary history of blood proteins, such as haemoglobin and serum albumin, in a great number of animals. Two men from the University of California at Berkeley, Vincent Sarich and Allan Wilson, have been applying one of those techniques—immunological reactions—to measuring molecular differences in serum albumin. And once again the results prove out, as the table above makes clear (the figures represent units of immunological difference measured against a standard that Sarich and Wilson use: sensitized rabbit serum albumin).

This table reveals some startling information. Not only does it confirm what other tests have shown —that man is very close to gorilla (only 8 differences), not so close to gibbon (14 differences) and even further from monkey (32 differences)—but it also shows that the monkey is equidistant from all the other three. This equidistance makes it possible to deduce that monkeys separated from the ancestor of all those apes at the same time, and that in all of them the rate of evolution in their serum albumin has been remarkably constant. In other words, they all have been evolving at about the same speed.

To check this all-important matter of rate of evolution, Sarich and Wilson went outside the primate family tree entirely and compared primates to carnivores. Their findings are shown in the second table. The number of serum-albumin changes here is higher, indicating a far more ancient split between primates and carnivores than among primates themselves. What is astonishing about these new figures is that, except for a slight difference in the tarsier, the figures are almost identical, proving again that these animals have all been evolving at the same rate.

The problem now is to determine that rate. For if we can measure the *amount* of evolution and the *rate* at which it is taking place, then we are back with the familiar time-rate-distance problems of our grade-school arithmetic books. With two variables known, we can calculate the third. Now, at last, we can measure evolutionary time accurately, and confidently begin marking down exactly when splits in a family tree took place.

Sarich and Wilson have addressed themselves to this task by assembling a great amount of molecular-biological evidence. By rigorous cross-checking of this material they arrived at tentative rates of evolution, not only for DNA but also for several blood proteins. Then they picked a date to start their primate family tree, selecting one on which some palaeontologists agree but many do not: a 36-million-year-old split between New World and Old World monkeys. With this as an anchor in time, they worked their way forward, using their molecular clocks to mark off splits as they went—monkey-ape first, and finally hominid-chimp.

What has brought the anthropological world to its feet with a roar of rage is that the Sarich-Wilson way of measuring things shows a hominid-chimp split that

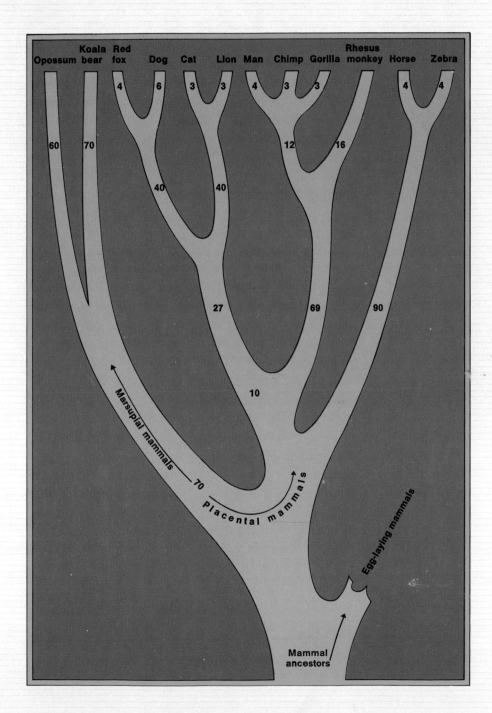

The labels in the diagram (left to right along the top): Opossum, Koala bear, Red fox, Dog, Cat, Lion, Man, Chimp, Gorilla, Rhesus monkey, Horse, Zebra

4, 6, 3, 3, 4, 3, 3, 4, 4

60, 70

12, 16

40, 40

27, 69, 90

10

Marsupial mammals

70

Placental mammals

Egg-laying mammals

Mammal ancestors

The Family Tree: Evidence in a Protein

The differences in the proteins of two species (*pages 132-133*) reflect evolution that has occurred in them since they split from a common ancestor. Analysis reveals that there are six differences between the protein serum-albumin of chimp and gorilla. This information dictates the first small fork (*top of diagram*) in this family tree devised by Vincent Sarich, molecular biologist and anthropologist at Berkeley. Man shows seven differences from both chimp and gorilla. Since the latter are already credited with three each, the remaining four go to man, and his fork can be plotted at about the same point on the tree as the chimp-gorilla fork.

A rhesus monkey was found to average 31 serum-albumin differences from the other three. This larger figure represents a greater degree of evolution, and thus an earlier split; accordingly, the fork separating monkeys from apes and men must be plotted lower down. Again, the laboratory technique shows continuing evolution in all lines, so the 31 differences must be divided, 16 for the monkey, 15 or 16 for the others. (For chimp, $3 + 12 = 15$; for man, $4 + 12 = 16$.)

Similar forks can be made for all the other animals shown. The horse and the zebra split eight differences between them. But each shows about 190 differences from primate serum-albumin, indicating an ancient split (about 94 for horse, about 96 for man).

is less than four million years old. Every bone man in the world is up in arms about that date. "What about Omo?" they shout. "Look at Kanapoi, at Lothagam. There are hominids from those places that are three, four and five million years old, and they don't look like apes. You're asking us to ignore fossils completely in favour of your precious molecules. You're asking us to assume constant rates of evolution, which we don't. And you're asking us to begin at a point in time that we cannot agree on among ourselves."

Here is a very serious dilemma. How can a science proceed that has palaeontologists insisting that the man-ape split took place at different times all the way back to 50 million years ago, and serologists insisting that it took place only four million years ago? Whom are we to believe?

Neither, suggests Sherwood Washburn. Although he is most impressed by the work that Sarich and Wilson are doing, he is also keeping a sharp eye on fossils. "*All* the clocks the immunologists have been using," he says, "contain various small errors. They run at different speeds, and for all we know, they may also be erratic. They have not yet been properly calibrated, but their hands are all pointing in the same general direction. That is very impressive."

What is needed, he goes on, is a way of matching the Sarich-Wilson data to a starting point in the geological-palaeontological past that all parties can agree on. Then, with some tugging and fitting, and with some adjustment of the clocks, all events might fall into place. If, for example, Sarich and Wilson had chosen a 50-million-year date for the New World/ Old World monkey split, then—on their scale of things—the man-ape split would have been pushed back beyond five million years ago. If, as some pa-

laeontologists wish, the monkey split were pushed still further back—say, to 75 million years—then the man-ape split would come at about seven or eight million years ago, and so on.

A man-ape split seven million years back, or—to stretch matters just a bit more—eight or nine million years back, begins to make the fossil problem less indigestible. It does not settle it. There is still the matter of how to regard 10- to 14-million-year-old Ramapithecus—known not to be an ape, strongly suspected of being a human forebear. But that problem should settle itself in time. More and better fossils could confirm what many now suspect—that Ramapithecus was a sort of early, perhaps-not-yet-erect, on-the-way-to-becoming Australopithecine. Or, they may confirm that Ramapithecus was not on the human line at all. That latter possibility will continue to exist until Ramapithecus is better known than it is today. And, of course, its credentials will vanish completely if an entirely new, and better, candidate for human ancestry is discovered.

But at the moment there is no such candidate. Ramapithecus is the best we have. Let us continue to consider it an ancestor, and attempt to trace our descent from there, incorporating the latest ideas that spring from the recent finds at Omo and East Rudolf.

Not long ago Bernard Campbell completed such an exercise: the last word in tree-making and hominid-naming. In the process he resolved two problems that I have deliberately left hanging. The first problem has to do with the relationship between the three Australopithecines: Boisei, Robustus, Africanus. The second deals with the problem of Habilis.

To recapitulate, the first discoveries of any Australopithecines were in South Africa and they were of two types, a gracile kind and a robust kind. Moving north to Olduvai, and later to Omo and Lake Rudolf, there are also two kinds. The smaller of these closely resembles the small southern gracile Africanus type, and is now believed by many to be the same species. But the big, husky northern one is so much more robust than its more modestly robust southern cousin that it has earned a separate name, Boisei, to distinguish it from the southern Robustus.

Looking at all three Australopithecines together, one is overwhelmed by the super-robustness of that massive-jawed northerner; by comparison, the differences between the two southern types begin to shrink rather dramatically. They are both so unlike Boisei that they begin to seem more like each other.

That explains the family tree on the next page, which diagrams Campbell's suggestion of how all three are related. He assumes an ancestral Australopithecine that descends from Ramapithecus. Some time between six and ten million years ago, this single species—following the familiar road of slow specialization to fit different niches—began to divide into two. One type, Boisei, found itself adapting more and more to a diet of rough vegetable matter. In time, this had a selective-adaptive effect on the teeth and jaws of that population, making incisors and canines small, but molars and jawbones increasingly large. The longest this specialization went on, the more pronounced it became. Since it seems to have gone on for several million years, that would explain the extreme massiveness of the jaws and back teeth of those very late Boisei fossils discovered by Richard Leakey at Lake Rudolf, and also the one Boisei skull discovered by his parents at Olduvai.

The gracile type, Africanus, meanwhile, found a different niche. Being more committed to an omnivorous diet, it was under less selective pressure to evolve large molars than it was to improve its meat-eating, hunting and tool-using proclivities—and to develop a generally more innovative and venturesome life style.

Gracile types may have been able to adapt better to slow climatic change, to live in a greater variety of environments, their habitat limited only by the availability of *some* kind of food and the existence of water within a few hours' walk. They probably spread much more widely over Africa than Boisei ever did, penetrating to the south, which is cooler, drier and subject to greater seasonal variation than the equatorial belt to the north.

It is significant that no Boisei fossils have ever been recovered from South Africa, and no Robustus fossils from the north. What are we to make of that peculiar distribution? Campbell suggests that the southern Robustus represents the beginnings of a late specialization of the gracile types that had penetrated South Africa. The reason they are less robust than Boisei is that the specialization down there took place more recently; robustness simply did not have as long a time to develop.

What Campbell is suggesting, as another look at his chart will show, is that while gracile types in the north were evolving into Habilis, and thence into Homo erectus, those in the south may have been following a scenario somewhat similar to that played out in the north several million years earlier. But this second time, the gracile population in the south did not continue on and evolve into Habilis—or at least, there is no fossil record that it did. Instead, it seems

The Family Tree: Fossils Reassessed

A family tree prepared in 1972 by Bernard Campbell of the University of California at Los Angeles includes insights derived from recent finds near the river Omo and Lake Rudolf, as well as from recent thinking about hominid types not included in the family trees on pages 135 and 140-141.

Campbell's scheme, more detailed than the others, depicts a vine-like interweaving of populations rather than simple branching (solid lines indicate species, dotted lines subspecies).

Campbell's tree presents Ramapithecus (*dark blue*) as the ancestor of Australopithecus africanus (*light blue*), and Boisei (*purple*) as a dead end. By the two-million-year mark, Africanus had split into Robustus, which became extinct, and Habilis (African) and Modjokertensis (found in Asia in 1936 and thought to be Homo erectus until recently), which gave rise to Erectus (*green*). Erectus developed geographical subspecies, from which several subspecies of Homo sapiens emerged (*yellow*).

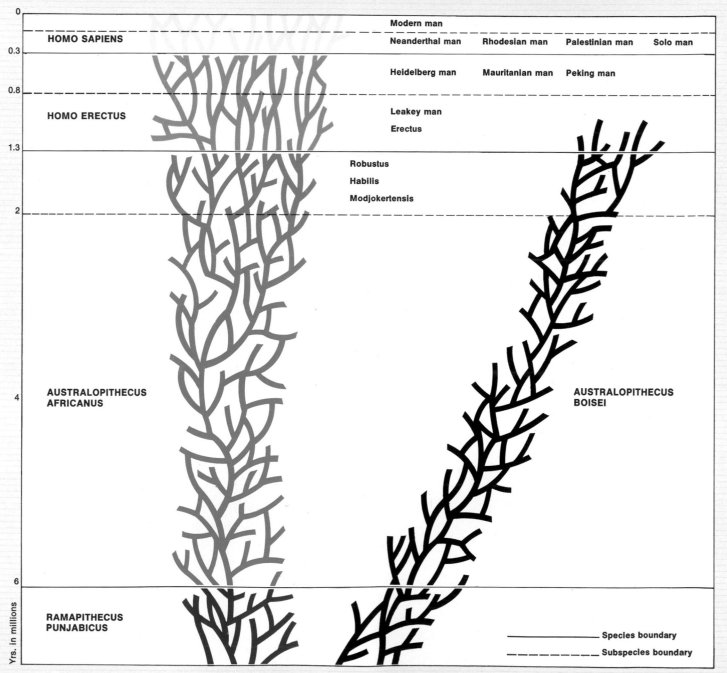

Yrs. in millions

0	
	Modern man
HOMO SAPIENS	Neanderthal man　　Rhodesian man　　Palestinian man　　Solo man
0.3	
	Heidelberg man　　Mauritanian man　　Peking man
0.8	
HOMO ERECTUS	Leakey man
	Erectus
1.3	
	Robustus
	Habilis
	Modjokertensis
2	

AUSTRALOPITHECUS AFRICANUS

AUSTRALOPITHECUS BOISEI

RAMAPITHECUS PUNJABICUS

————————— Species boundary

— — — — — — Subspecies boundary

to have evolved into Robustus. Despite the lack of sound dating in South Africa, it is generally agreed, on other evidence, that the robust fossils down there are all much younger—on the order of at least a million years younger—than the gracile ones.

Campbell deals with this tricky problem by allowing for the evolution in different places of different populations of Africanus. In the north it becomes Habilis. In the south it becomes Robustus.

And there may even be a third line of Australopithecines that developed in Asia. The picture of that third line is extremely confused. It goes back to the 1930s and to some fossils found in Java by the Dutch anthropologist G. H. R. von Koenigswald. For some years those were generally regarded as primitive examples of an Asiatic race of early Homo erectus types that were ancestral to the Java ape-man, Peking man and others. Now, with the availability of Habilis fossils from Africa, it is possible to compare them with Von Koenigswald's finds and to detect characteristics that they have in common. Is Von Koenigswald's type a Habilis also? We don't really know. The fossil evidence is meagre; dating is uncertain; some of the original sites have been disturbed to the point where they are no longer useful. Nevertheless, Campbell finds enough Habilis traits in the Von Koenigswald type (which bears the jaw-breaking name of *Modjokertensis*) to include it in his family tree as yet a second Africanus-into-Homo.

Africanus, then, must have been a progressive, adaptable and successful creature, spreading widely through the tropics of the Old World and, like any such creature, evolving as it went. This is an old page out of the evolutionary script: three populations, left to themselves long enough, turning out differently.

At least that is the way Campbell proposes it.

The Campbell scenario also disposes of the problem of how to regard Habilis. The question left hanging about him was: is he just a late Africanus or is he a separate species on his own? According to Campbell, he is the former, a very late Africanus indeed, showing up at about the two-million-year mark and submerging himself into Erectus at approximately 1.3 million years ago.

Could Campbell be wrong? Certainly he could. He would be the first to admit this. What he is doing —along with everybody else who is engaged in putting together this fascinating jigsaw puzzle—is rearranging what evidence he has in the light of new evidence as fast as it appears. The differences between Boisei, Africanus and Robustus compel some kind of a diagrammatic accommodation. Campbell has supplied a logical one. Not everybody will agree with it, notably those who still believe that all Australopithecine forms are varieties of a single species, and those who still insist that the two southern types are merely males and females of the same species.

On that note the case for man's origin as an Australopithecine must rest for the time being. And the summing up to the jury might run something like this:

Australopithecines are descended from apes. Studies of fossils, and studies of both the physical nature and the behaviour of living apes, make this clear. Their closest ape relatives were chimpanzee and gorilla. They followed a different evolutionary course from those apes by invading a niche that neither the chimpanzee nor the gorilla occupied: a life on the open ground. This led—through complex feedback interactions of manual dexterity, tool use, bipedalism and hunting—to a way of life that came to include the

The Family Tree: A Spectrum of Views

These family trees show how differently five experts interpret the evidence—and how theories change with new finds. The trees are plotted to the same geologic time periods, but no dates are given as the experts do not agree on the time spans. Colours key

hominids to the chart on page 138.

The first tree was made by the late Sir W. Le Gros Clark, the famed British palaeontologist, in 1959. It lists a number of extinct apes at the left, one line leading to modern gibbons. Clark put Australopithecus on a dead end, as he did Homo erectus, Neanderthal man and Rhodesian man.

The tree published in 1971 by John

Napier of Queen Elizabeth College, London, starts with a very ancient primate, Aegyptopithecus, and leads to Ramapithecus. The earliest Homo, Napier calls Habilis, later ones Erectus. He recognizes only two Australopithecine types, Africanus and Boisei (which he calls Paranthropus), and considers neither ancestral to man.

Focusing on hominids, Phillip J. To-

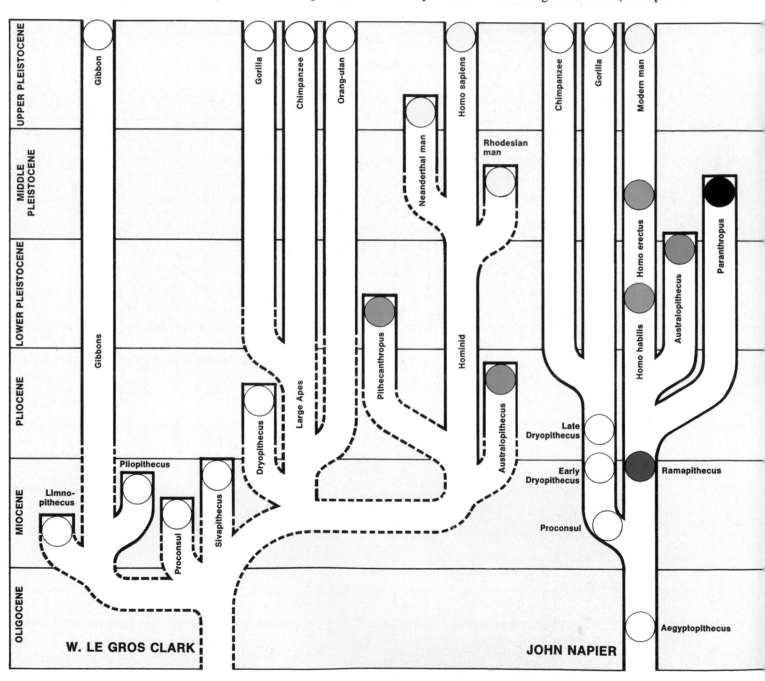

bias of the University of Witwatersrand, South Africa, starts his 1965 tree with an ancestral Australopithecine. Boisei and then Africanus branch off and become extinct, while the ancestral Australopithecine produces Homo habilis, Homo erectus, Neanderthal and modern man.

C. Loring Brace of the University of Michigan holds a simple view of hominid evolution. His 1971 tree runs from Aegyptopithecus straight through to modern man. He recognizes only one form of Australopithecus.

The fifth tree, also drawn by Brace in 1971, reflects the views of Louis Leakey, discoverer of the fossils of Olduvai Gorge. Here, Proconsul (which some consider closer to a gorilla than a hominid) is on the main stem. Leak-ey called Ramapithecus Kenyapithecus, and recognized but one Australopithecine, the super-robust Boisei. What some call Africanus or Australopithecus habilis he named Homo habilis, which leads directly to modern man, with a possible offshoot towards a line that contains Pithecanthropus (the Java specimen of Homo erectus) and ends with Neanderthal.

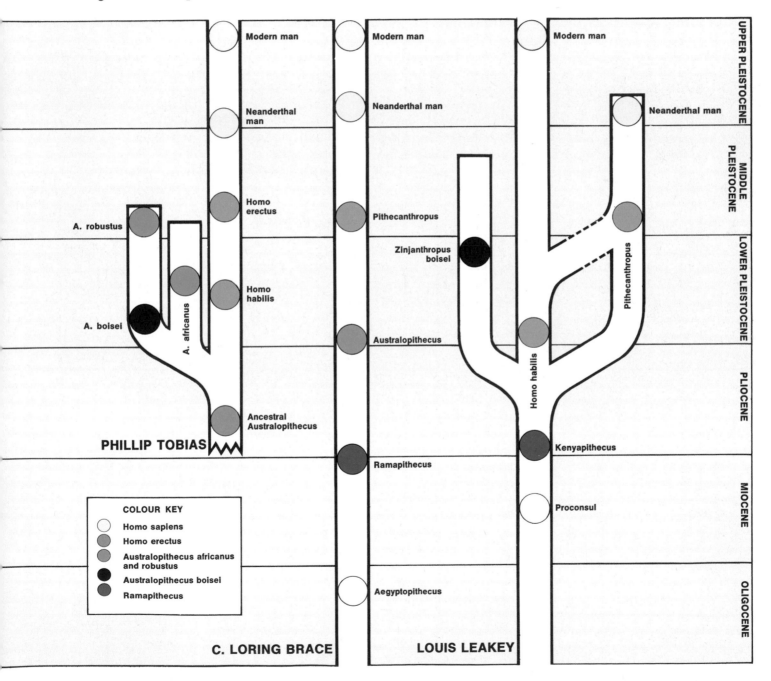

COLOUR KEY

- Homo sapiens
- Homo erectus
- Australopithecus africanus and robustus
- Australopithecus boisei
- Ramapithecus

PHILLIP TOBIAS

C. LORING BRACE

LOUIS LEAKEY

making of tools, food sharing, family formation, ever-increasing intelligence—finally reaching a stage of development that permits them to be called men. That stage began to be reached some two million years ago, and had definitely arrived 1.3 million years ago.

This leaves a time span of approximately 700,000 years—give or take a couple of hundred thousand —during which progress towards the human estate continued to be made. Progress was probably not steady, nor is there any reason to suppose that it took place at the same rate in different parts of the world. In fact, it is highly unlikely that it did, since organisms adapt to environments, and environments the world over are different. There may well have been isolated populations here and there that simply died out. If so, their fossils, when discovered, will provoke some head-scratching—as some pockets of peculiar, isolated, late-staying Neanderthals from north-west Europe have done.

However that may be, the necessary ingredient for overall evolution in the species that was Australopithecus and became Homo was the maintenance of contact—no matter how remote or slender—to ensure the continuing exchange of genes.

That is the lesson that the wiggly little lines on the Campbell family tree teach us. Evolution is like a vine, with many tendrils branching out, some to wither and die, but most of them curling back to recombine with other tendrils and form a multi-stranded column, not a single solid trunk.

It is that model of the family tree that we must keep in mind, and it is that critical 700,000-year period that we must keep our eyes on. If there is a missing link—anything that can justifiably be called one—in that series of links and tendrils that make up the chain of hominid evolution, then the link must be that ex-Homo habilis, an Australopithecine-becoming-man who lived on the African savanna, close to the streams and the lake edges of an ancient landscape that was essentially no different from similar landscapes found in Africa today. Somewhere on that distant shore—and extending over a period of about three-quarters of a million years—is the borderland between apehood and humanity.

The Experts Have Their Say

The 15 authorities shown here are identified on the following pages, along with their often conflicting views about how man evolved.

Palaeoanthropology has been the most argumentative of sciences since its beginning. Experts who agree are rare.

In the early days of the science the supply of fossil material was so meagre that one man's theory about how hominid skulls, teeth and jaws were related was as good as another's. There simply was not enough evidence on hand to substantiate or refute either one. Today the situation is entirely different. The supply of fossils has become a small flood, flowing into museum and laboratory almost faster than it can be studied. New disciplines have been drawn in to help analyse these fossil finds: geology, physics, botany, climatology, chemistry, animal behaviour, molecular biology. Each contributes fresh insights, but each tends to create fresh problems at the same time because these new disciplines do not always agree. Never has there been a greater whirl of informed, persuasive—and conflicting—theory about man's origins.

The people shown above are 15 scientists who made major contributions to the debate over human evolution in the early 1970s. Each contributed something to the ideas expressed in this book, for each took his own view of the emergence of man.

SHERWOOD L. WASHBURN

A professor at the University of California of Berkeley, Washburn is one of the world's foremost authorities on primate behaviour. But he believes that a great many disciplines besides his own are vital to an understanding of human evolution, and he invites any number of players to join in the sport.

The study of human evolution is a game, rather than a science in the usual sense. The remote past cannot be brought into the laboratory and subjected to carefully controlled experiments. In spite of all the recent advances in the understanding of the mechanics of evolution and of primate fossils, scientists still disagree. Some think that man has been separated from other primates for 50 million years; others for less than one fifth of that.

My interests have been in learning to play the evolution game. A bone may seem a relatively simple object, one that is easy to describe and interpret. Yet, when the bone was important it was part of a living animal, and it is amazing how different the bone appears after one has been watching living monkeys and apes.

LOUIS LEAKEY

Right up to his death in 1972, Louis Leakey was one of the most controversial figures in palaeoanthropology, pressing arguments peculiarly his own. His fossil finds were varied and legendary—including the extremely ancient creature, Kenyapithecus africanus, which he considered a hominid, from rock layers 20 million years old, as well as 14-million-year-old fossils of the remote human ancestor, Ramapithecus. From such discoveries he constructed a family tree (page 141) that shows Australopithecines as cousins, not direct ancestors, of the human genus, Homo.

The Australopithecines developed away from true man (Homo), who was approximately contemporary. By three million years ago, both forms were present in East Africa.

My finds show that man's ancestral stock separated from that of the great apes more than 20 million years ago. They also show that the genus Homo dates back in East Africa 1.5 to three million years, that a form of Homo erectus was present in Africa before Asia and finally, that "near man", Australopithecus, was developing in parallel and died out 50,000 to 1.5 million years ago.

The time span of psycho-social man —40,000 years—represents a moment compared with the 20 million years of hominid existence. We can, therefore, expect a long future ahead if we do not destroy ourselves and the world.

DIAN FOSSEY

Miss Fossey is an authority on the behaviour of mountain gorillas in central Africa, the same terrain that their ancestors and Australopithecus lived on before the man-ape left for the open plains. She finds parallels between gorilla and human social behaviour, and these similarities suggest to her that Australopithecus may, like the modern gorilla, have been a nonaggressive creature willing to share his feeding grounds with his fellows.

An expert in the detailed measurement and study of Australopithecine fossils, Tobias has probably handled more of these specimens than any other man alive. He considers himself "one of the fathers" of Homo habilis, perhaps the most argued-over member of man's family tree. He views Habilis as a very early true human, while to others Habilis is an advanced Australopithecine—perhaps the missing link.

RAYMOND DART

In 1924 Raymond Dart startled the anthropological world by his discovery of a skull fossil in a rock quarry at Taung, South Africa. He named it Australopithecus africanus and boldly declared it to be a human ancestor. His claim was derided. But the profusion of subsequent finds at Taung and elsewhere, including a wealth of tools among the bones, proved him right and led him to conclude that Australopithecus walked upright—and was a cannibal.

Gorillas have an extremely cohesive social structure. Like the early terrestrial hominids, they travel, sleep and feed as a group. Initially Africanus and Robustus groups may have been able to live side by side, their food needs varying enough so that they could share overlapping areas and avoid a sense of territoriality. The question then arises; did Robustus groups defend their homegrounds against other Robustus groups? Gorilla groups, though adhering to limited ranges, lack a strong sense of territoriality, due to the even distribution of foliage growth within their environment. Thus they overlap and share range areas.

Our claim that Homo habilis was a species contemporary with Australopithecus aroused nearly as much opposition as did the first recognition of Australopithecus. But my recent demonstrations that the brain size of Habilis specimens from East Africa is half again as large as the average brain size of Australopithecus provide strong confirmation that we are dealing with something more advanced in the human direction. I am not terribly worried whether the name Homo habilis stands the test of time. What is important is that it represents a population of early hominids bigger in brain size and more certainly related to stone cultures than any of its contemporaries or predecessors. It stands four-square on the road to man.

When in 1924 I identified the Taung fossil Australopithecus africanus, I based its ancestral significance on characteristics of the fossil skull and on inferred features related to biped walking. But my analysis of damaged baboon skulls demonstrated the Australopithecines' predatory and cannibalistic habits and usage of long bones as bludgeons and severed skulls as receptacles. I do not accept the idea that the Australopithecines had speech. Their familial hunting horde-life needed little more communication than other predators.

DAVID PILBEAM

Pilbeam is one of the world's authorities on Australopithecine fossils and on the fossils of the Dryopithecine apes that many believe preceded them. He is willing to place odds of three to one that hominids diverged from apes in the very distant past, 15 million years ago.

I think Ramapithecus species of Africa and India are hominids. I have grown increasingly sceptical of the view that hominids differentiated as weapon-wielding savanna bipeds. I am as inclined to think that changes in a predominantly vegetarian diet provided the initial impetus. Also I believe that too little emphasis has been placed on the rôle of language and communication, and too much on tools, in understanding the later stages of human evolution.

J. T. ROBINSON

Zoologist Robinson has to his credit more than 300 fossil discoveries in Sterkfontein, Kromdraai and Swartkrans, in South Africa. His studies of the large and small—robust and gracile—forms of Australopithecus lead him to conclusions totally different from those drawn by Richard Leakey (opposite), and his father, Louis (page 144).

The two large samples of hominid from Sterkfontein and Swartkrans provide excellent evidence against the view that the robust and gracile forms are males and females of the same species. On this hypothesis the males would have had canine teeth proportionately much smaller than the females—a feature not known in any higher primate. Also the females would have been efficiently striding, erect bipeds, very similar in this respect to modern man, while the males would have lacked the ability to stride. Such a population makes very little biological sense.

F. CLARK HOWELL

Howell is known for the painstaking methods he has developed for excavating the remains of prehistoric creatures. Thanks to such methods, the hominid finds in East Africa could be compared with those from other sites, and the significance of the very old Ramapithecus fossils could be recognized. Only when reliable data of this kind became available could scientists begin to zero in on the fundamental questions about the emergence of man: when and where did his ancestors originate?

In recent years there has been increasing concern with the nature of the source and time of origin of the hominids. The problem was posed, of course, by the discovery and recognition of the Australopithecines. But it has only been possible to consider this question thoughtfully after the investigations at Olduvai, Omo, East Rudolf, Kanapoi and Lothagam, and the recognition of the significance of Ramapithecus from India and Kenya.

We still do not know the source of the hominids, but it is possible that their origin may lie between 7 and 15 million years ago, and perhaps not only in Africa. This time range is still not well known. Anyone who feels that we already have the problem solved is surely deluding himself.

RICHARD LEAKEY

While still in his twenties, he built a formidable reputation for himself with his discoveries in East Rudolf—including a skull in 1972 he considers truly human but 2.5 million years old, far more ancient than any other. They support controversial ideas about man's origin—its very ancient date and the rôle of tool use in the development of human features (pages 49 and 50), and the possibility that the great disparity in size among Australopithecus specimens may merely represent the difference between male and female forms of one species.

In East Rudolf we found a vast site covering close to a thousand square miles, with lake beds spanning a period from about five million to just under one million years ago. Three benefits have come from the work there. First, we now have enough reasonably complete specimens to begin to discuss chewing mechanisms, and to begin to interpret patterns of locomotion. Second, we have documented that Australopithecus boisei included very large males and much smaller females, which, had they been found separately, might have been called separate species.

Finally, we have clear evidence of the coexistence of Australopithecus and Homo. This means that the genus Homo didn't evolve from Australopithecus within the last million years, but that both arose from the same stock some four to five million years back.

Boisei, two million years ago, was a specialized herbivore. I do not think he was fully upright, and I also think it is a mistake to consider Australopithecine locomotion as intermediate between quadruped and biped. I think it represents something unique that became extinct.

Homo, on the evidence from Rudolf, was certainly upright. Early man was a hunter, but I think the concept of aggressiveness—the killer-ape syndrome —is wrong. I am quite sure that the willingness of modern aggressive man to kill his own kind is a very recent cultural development, probably linked to material-based society, fixed settlement, property and so on.

GEORGE B. SCHALLER

Schaller used to study one of man's primate cousins, the mountain gorilla, for clues to human behaviour but some years ago shifted his search to the habits of animals quite unrelated to man—social carnivores, which eat meat and hunt for it in packs. Here he tells why.

Man is a primate by inheritance but a carnivore by profession. If we are to understand the evolutionary forces that shaped his body, mind and society, it is essential that we recognize his dual past.

It has become clear that social systems are so strongly influenced by ecological conditions that similarities in the societies of human primates and man might have occurred by chance. Monkeys and apes are essentially vegetarians, and confine their existence to small ranges. Man, on the other hand, has been a widely roaming hunter and scavenger for over two million years. Species that are genetically unrelated but ecologically similar to early man—such as the African hunting dog and the lion—might teach us more than any nonhuman primate could about the evolutionary pressures that shaped and maintained our society. African hunting dogs, for instance, have a division of labour. Some adults guard pups while others search for prey: they also hunt co-operatively and share food equally, traits that are not prominent among nonhuman primates but that are thought to have been important in the genesis of human society.

JANE GOODALL

Most of her professional life has been lived in the Gombe Stream Reserve in western Tanzania, where she studies the chimpanzee. It is man's closest living relative—both in physical and behavioural terms—and it, Jane Goodall thinks, is the key to man's understanding of himself.

The Gombe chimpanzees have been observed to use and make simple tools for feeding, cleaning themselves, investigating their environment and as weapons. They frequently hunt small animals for food and may show quite elaborate group hunting techniques. The affectionate bond between mother and offspring and between siblings is extremely strong and may persist through life.

The similarities of some nonverbal communication patterns in man and chimpanzee are striking; not only may gestures themselves be similar but also the contexts in which they are likely to occur. Chimpanzees and humans, when greeting after separation, may embrace, kiss, pat each other or clasp hands.

It is my belief that a better understanding of the behaviour of man's closest living relative will suggest new lines of inquiry into the biological basis of certain aspects of human behaviour—especially problems connected with child-raising, adolescence, aggression and some mental disorders.

MARY LEAKEY

It was Mary Leakey—wife of Louis and mother of Richard—who made the first discovery of an Australopithecus fossil at Olduvai Gorge. She has since concentrated on stone tools and is now a world authority on the subject.

I am not a physical anthropologist and prefer not to hazard a guess as to what may have been man's earliest ancestor. In fact, I prefer the criterion for "man" to be based on the establishment of organized tool-making rather than on the morphology of any particular fossil.

Man's ancestors must inevitably have passed through a stage without even the conception of tools, to a stage in which natural objects were used for various purposes, and from there to a stage in which the natural objects were modified by means of the hands and teeth. Finally, one tool was used to make another, i.e., a hammerstone for detaching flakes from a chopper. This, for me, is the stage to which we can apply the term Homo.

I do not really believe that the study of fossil man contributes to any measurable degree to the understanding of modern man.

BERNARD CAMPBELL

Campbell is a physical anthropologist —one of that group who can tell from a jawbone (or even only a tooth) what kind of food a creature lived on, from foot bones how it walked and from hand bones what kind of tools it was able to make. With these observations he integrates knowledge from such disciplines as ecology, psychology and field studies of animal behaviour to attempt to shed light on problems that have troubled philosophers for millennia—the sources of violence, the meaning of love, the power of family ties.

My central research interest has been in the interpretation of fragments of fossil hominids—to understand their anatomy and, as a result, not only their ecological context but their behaviour and their whole way of living as well. Modern man is a product of his genes and his environment, and a full understanding of human behaviour requires knowledge of the genetic roots of man's behaviour as well as the social environment into which each individual is born, grows up and matures. Studies of human prehistory seem to me to be of central importance in understanding ourselves, a challenge whose solution can mean life or death to the human species.

G. H. R. VON KOENIGSWALD

Von Koenigswald found human and Australopithecine fragments at Sangiran in Java remarkably like the fossils of Olduvai in Africa, nearly 5,000 miles away. That discovery, together with a find in India of Ramapithecus, a presumed ancestor of Australopithecus, led him to place the site of man's origin far from the regions favoured by others.

I definitely believe man's earliest ancestors came from Asia, where Ramapithecus lived about 10 million years ago. In Middle Java there were remains of primitive man (Pithecanthropus) as well as Australopithecines (Meganthropus) alongside each other.
  This duplication is curious indeed: it means that similar situations existed on both sides of the Indian Ocean, both in Olduvai and Sangiran. Geographically seen, the distance of Java from India and of Olduvai from India is about the same. This raises the possibility that human development originated in India.

VINCENT SARICH

Sarich sees the search for answers to the puzzle of evolution as a game, as does his colleague Sherwood Washburn (page 144). But Sarich plays a different position on the team; he is a molecular biologist who constructs family trees by protein analysis. Here he explains how that new approach fits into the scheme.

What we want is an evolutionary history. As long as there was only the fossil-based picture, one could disagree with that picture only by arguing about how to interpret anatomical data—and such arguments have been going on for a hundred years. What the molecules provide is a new set of rules that limits the possible interpretations. The biochemist knows his molecules had ancestors —while the palaeontologist can only hope his fossils had descendants.

The Emergence of Man

This chart records the progression of life on earth from its first appearance in the waters of the newly-formed planet to the evolution of man; it traces his physical, social, technological and intellectual development to the Christian era. To place these advances in commonly used chronological sequences, the column

Geology	Archaeology	Thousand Millions of Years Ago	
Precambrian earliest era		4.5	Creation of the Earth
		4	Formation of the primordial sea
		3	First life, single-celled algae and bacteria, appears in water
		2	
		1	

Geology	Archaeology	Millions of Years Ago	
			First oxygen-breathing animals appear
		800	
Palaeozoic ancient life			Primitive organisms develop interdependent specialized cells
		600	Shell-bearing multicelled invertebrate animals appear
			Evolution of armoured fish, first animals to possess backbones
		400	Small amphibians venture on to land
			Reptiles and insects arise
			Thecodont, ancestor of dinosaurs, arises
Mesozoic middle life		200	Age of dinosaurs begins
			Birds appear
			Mammals live in shadow of dinosaurs
			Age of dinosaurs ends
		80	
			Prosimians, earliest primates, develop in trees
Cainozoic recent life		60	
		40	**Monkeys and apes evolve**
		20	
		10	**Ramapithecus, oldest known primate with apparently man-like traits, evolves in India and Africa**
		8	
		6	
		4	**Australopithecus, closest primate ancestor to man, appears in Africa**

Geology	Archaeology	Millions of Years Ago	
Lower Pleistocene oldest period of most recent epoch	**Lower Palaeolithic** oldest period of Old Stone Age	2	**Oldest known tool fashioned by man in Africa**
			First true man, Homo erectus, emerges in East Indies and Africa
		1	Homo erectus migrates throughout Old World tropics

		Thousands of Years Ago	
Middle Pleistocene middle period of most recent epoch		800	Homo erectus populates temperate zones
			Man learns to control and use fire
		600	
			Large-scale, organized elephant hunts staged in Europe
		400	Man begins to make artificial shelters from branches
		200	
Upper Pleistocene latest period of most recent epoch	**Middle Palaeolithic** middle period of Old Stone Age		Neanderthal man emerges in Europe
		80	
		60	Ritual burials in Europe and Middle East suggest belief in afterlife
			Woolly mammoths hunted by Neanderthal in northern Europe
		40	Cave bear becomes focus of cult in Europe
	Upper Palaeolithic latest period of Old Stone Age	30	Cro-Magnon man arises in Europe
			Man reaches Australia
			Oldest known written record, a lunar calendar on bone, made in Europe
			Asian hunters cross Bering Strait to populate North and South America
			Figurines sculpted for nature worship
			First artists decorate walls and ceilings of caves in France and Spain
		20	Invention of needle makes sewing possible
			Bison hunting begins on Great Plains of North America
Holocene present epoch	**Mesolithic** Middle Stone Age	10	Bow and arrow invented in Europe
			Dog domesticated in North America

(Last Ice Age — spanning Upper Pleistocene through Upper Palaeolithic)

Four thousand million years ago — Three thousand million years ago

Origin of the Earth (4,500 million) — First life (3,500 million)

he far left of each of the chart's four sections identifies the great
ogical eras into which earth history is divided, while the second
mn lists the archaeological ages of human history. The key dates
he rise of life and of man's outstanding accomplishments appear
he third column (years and events mentioned in this volume of

The Emergence of Man appear in bold type). The chart is not to
scale; the reason is made clear by the bar below, which represents
in linear scale the 4,500 million years spanned by the chart—on the
scaled bar, the portion relating to the total period of known human
existence (far right) is too small to be distinguished.

eology	Archaeology	Years B.C.	
locene (ont.)	Mesolithic (cont.)	9000	Jericho settled as the first town
			Sheep domesticated in Middle East
	Neolithic New Stone Age		
		8000	Pottery first made in Japan
			Goat domesticated in Persia
			Man cultivates his first crops, wheat and barley, in Middle East
		7000	Pattern of village life grows in Middle East
			Catal Huyuk, in what is now Turkey, becomes the first trading centre
			Loom invented in Middle East
			Agriculture begins to replace hunting in Europe
		6000	Cattle domesticated in Middle East
	Copper Age		Copper used in trade in Mediterranean area
			Corn cultivated in Mexico
		4000	Sail-propelled boats used in Egypt
			Oldest known massive stone monument built in Brittany
			First cities rise on plains of Sumer
			Cylinder seals begin to be used as marks of identification in Middle East
		3500	First potatoes grown in South America
			Wheel originates in Sumer
			Egyptian merchant trading ships start to ply the Mediterranean
			First writing, pictographic, composed, Middle East
	Bronze Age	3000	Bronze first used to make tools in Middle East
			City life spreads to Nile Valley
			Plough is developed in Middle East
			Accurate calendar based on stellar observation devised in Egypt
			Sumerians invent potter's wheel
			Silk moth domesticated in China
			Minoan navigators begin to venture into seas beyond the Mediterranean
		2600	Variety of gods and heroes glorified in *Gilgamesh* and other epics in Middle East
			Pyramids built in Egypt
		2500	Cities rise in the Indus Valley

Geology	Archaeology	Years B.C.	
Holocene (cont.)	Bronze Age (cont.)	2400	Stonehenge, most famous of ancient stone monuments, begun in England
			Earliest written code of laws drawn up in Sumer
		2000	Chicken and elephant domesticated in Indus Valley
			Use of bronze spreads to Europe
			Eskimo culture begins in Bering Strait area
			Man begins to cultivate rice in Far East
			Herdsmen of Central Asia learn to tame and ride horses
		1500	Invention of ocean-going outrigger canoes enables man to reach islands of South Pacific
			Oldest known paved roads built in Crete
			Ceremonial bronze sculptures created in China
			Imperial government, ruling distant provinces, established by Hittites
		1400	Iron in use in Middle East
	Iron Age		First complete alphabet devised in script of the Ugarit people in Syria
			Hebrews introduce concept of monotheism
		1000	Reindeer domesticated in northern Europe
		900	Phoenicians develop modern alphabet
		800	Celtic culture begins to spread use of iron throughout Europe
			Nomads create a far-flung society based on the horse in Russian steppes
			First highway system built in Assyria
			Homer composes *Iliad* and *Odyssey*
		700	Rome founded
			Wheelbarrow invented in China
		200	Epics about India's gods and heroes, the *Mahabharata* and *Ramayana*, written
			Water wheel invented in Middle East
		0	Christian era begins

▼ Two thousand million years ago   ▼ One thousand million years ago

First oxygen-breathing animals (900 million) ▲   First animals to possess backbones (470 million) ▲   First men (1.3 million) ▲

Credits

The sources for the illustrations in this book are shown below. Credits from left to right are separated by semicolons, from top to bottom by dashes.

Cover—Painting by Herb Steinberg, background photograph by Dr. Edward S. Ross. 8 —Painting by Burt Silverman, background photograph by Alfred Eisenstaedt for LIFE. 13—Map by Adolph E. Brotman. 16,17—Paper sculpture by Nicholas Fasciano, photographed by Ken Kay. 21 to 31—Paintings by Burt Silverman, background photographs are listed separately: 21—Pete Turner. 22,23—J. Alex Langley from D.P.I.; Emil Schulthess from Black Star. 24,25—Pete Turner. 26,27—Maitland A. Edey. 28,29—Dale A. Zimmerman and Marian Zimmerman; Maitland A. Edey. 30,31—Constance Hess from Animals Animals. 32—Michael Irwin courtesy Transvaal Museum, Pretoria, South Africa. 35—Fritz Goro, Peabody Museum of Natural History, Yale University. 36—Drawing by Adolph E. Brotman. 40,41—Fritz Goro, Peabody Museum of Natural History, Yale University. 42,43—Drawings by Susan Fox. 50—Fritz Goro, Museum of Comparative Zoology, Harvard University. 54 to 62—Paintings by Don Punchatz. 67—John Reader for LIFE. 68—Gordon W. Gahan, National Geographic Society. 69—Dr. Roger C. Wood. 70,71—John Reader for LIFE. 72,73 —Dr. Roger C. Wood. 74,75—Gerald G. Eck. 76,77—Frank Woehr from Photo Trends. 78 —Hugo van Lawick, National Geographic Society. 82,83—Designed by Jeheber & Peace, Inc. Illustrations by Robert Frost. 88—Hugo van Lawick, National Geographic Society. 90 to 95—Dr. Timothy W. Ransom. 98—Hugo van Lawick, National Geographic Society. 99—Dr. Timothy W. Ransom, National Geographic Society—Patrick P. McGinnis, National Geographic Society. 102 —Dr. Timothy W. Ransom, National Geographic Society. 106—Maitland A. Edey. 110, 111,112—Drawings by Nicholas Fasciano. 113—Alan Root. 114,115—Drawing by Nicholas Fasciano; Willard Price. 116—Drawing by Nicholas Fasciano. 117—Alan Root except top left, Hugo van Lawick. 122 to 125 —John Reader courtesy National Museums of Kenya. 128—Leonard Wolfe courtesy Carnegie Institution, Washington, D.C. 132—Designed by Jeheber & Peace, Inc. Illustrations by Robert Frost. 135 to 141—Drawings by Adolph E. Brotman. 143—Credits for this page appear on pages 144 to 149: 144 —Brian L. O'Connor; Gordon W. Gahan, National Geographic Society. 145—Photograph by Robert M. Campbell © National Geographic Society; Michael Irwin; Dr. C. K. Brain. 146—Cynthia Ellis; Margaret E. Donnelly; Ted Streshinsky. 147—John Reader for LIFE; Kay Schaller. 148—From *In the Shadow of Man*, by Jane van Lawick-Goodall. Photographs by Hugo van Lawick. Copyright © 1971 by Hugo and Jane van Lawick-Goodall. Reprinted by permission of Houghton Mifflin Company; Gordon W. Gahan, National Geographic Society. 149—Enrico Ferorelli; Courtesy Professor G. H. R. von Koenigswald; Dr. George Mross.

Acknowledgments

Parts of this book were read, with much helpful criticism and suggestion, by the following: John Crook, Professor of Psychology, The University, Bristol, England (on baboon behaviour and social organization); Jane Goodall, Scientific Director, Gombe Stream Reserve Research Centre, Kigoma, Tanzania (on chimpanzee behaviour); F. Clark Howell, Professor of Anthropology, University of California at Berkeley (on Omo fossil finds); Clifford J. Jolly, Associate Professor of Anthropology, New York University (on primate behaviour, the evolution of hominids as seed eaters and fossil interpretations); Mary D. Leakey, Leader, Olduvai Gorge Research Project, Langata, Nairobi, Kenya (on hominid tools and tool making); Richard Leakey, Director, National Museum of Kenya, Nairobi (on East Rudolf fossil finds); David Pilbeam, Associate Professor of Anthropology, Yale University (on a general review of Australopithecine and pre-Australopithecine fossils and dates); Vincent M. Sarich, Associate Professor of Anthropology, University of California at Berkeley (on molecular biological evidence in evolutionary studies and dating); and George B. Schaller, Research Associate, New York Zoological Society and Rockefeller University's Institute for Research in Animal Behavior, New York City (on gorilla behaviour and social carnivores as models for hominids as hunters).

The author and editors also wish to thank the following: Kay Behrensmeyer, Museum of Comparative Zoology, Harvard University; Edward Berger, Research Scientist, Special Research Laboratory, Veterans Administration Hospital, New York City; Claud Bramlett, Assistant Professor of Anthropology, University of Texas; Raymond A. Dart, Emeritus Professor, University of Witwatersrand, Johannesburg, South Africa; Phyllis Jay Dolhinow, Associate Professor of Anthropology, University of California at Berkeley; Gerald Eck, Department of Anthropology, University of California at Berkeley; Rhodes W. Fairbridge, Professor of Geology, Columbia University; Dian Fossey, Ruhengeri, Rwanda, East Africa; David Hamburg, Chairman, Department of Psychiatry, Stanford University School of Medicine; B. H. Hoyer, Carnegie Institution, Washington, D.C.; Glynn L. Isaac, Associate Professor of Anthropology, University of California at Berkeley; Richard F. Kay, Peabody Museum of Natural History, Yale University; L. S. B. Leakey, Senior Pre-Historian, National Museum of Kenya, Nairobi; Bryan Patterson, Professor of Vertebrate Paleontology, Harvard University; Timothy W. Ransom, University of California at Berkeley; Nancy Rice, Carnegie Institution, Washington, D.C.; John T. Robinson, Professor of Zoology, University of Wisconsin, Madison; Elwyn L. Simons, Professor of Vertebrate Paleobiology and Primatology, Peabody Museum of Natural History, Yale University; Richard H. Tedford, Curator, Department of Vertebrate Paleontology, American Museum of Natural History, New York City; Phillip V. Tobias, Department of Anatomy, Faculty of Medicine, University of Witwatersrand, Johannesburg, South Africa; Ralph von Koenigswald, Senckenberg Museum, Frankfurt am Main, Germany; Adrienne Zihlman, Assistant Professor of Anthropology, University of California at Santa Cruz.

Bibliography

General

Campbell, Bernard, *Human Evolution*. Heinemann, 1967.

"Conceptual Progress in Physical Anthropology—Fossil Man." *Annual Review of Anthropology* (Vol. I), 1972.

Clark, John Desmond, *The Prehistory of Africa*. Thames and Hudson, 1970.

Darwin, Charles, *On the Origin of Species*. Oxford University Press, 1951.

Ehrlich, Paul, and Richard Holm, eds., *Process of Evolution*. McGraw Hill, 1963.

Kurten, Bjorn, *Age of Mammals*. Weidenfeld and Nicolson, 1971.

Le Gros Clark, Wilfred E., *Man-Apes or Ape-Men?* Holt, Rinehart and Winston, 1967.
The Antecedents of Man. British Museum, 1970.

Leakey, Louis S. B., *The Progress and Evolution of Man in Africa*. Oxford University Press, 1961.

Leakey, Louis S. B., and Eve Goodall, *Unveiling Man's Origins*. Methuen, 1970.

Morgan, Elaine, *The Descent of Woman*. Souvenir Press, 1972.

Napier, John, *The Roots of Mankind*. Allen and Unwin, 1971.

Pfeiffer, John, *The Emergence of Man*. Nelson, 1970.

Pilbeam, David, *The Evolution of Man*. Thames and Hudson, 1970.

Washburn, Sherwood L., *Classification and Human Evolution*. Methuen, 1964.

Washburn, Sherwood L., and Phyllis C. Jay, eds., *Perspectives on Human Evolution*. Holt, Rinehart and Winston, 1968.

Carnivores

Matthiessen, Peter, and Eliot Porter, *The Tree Where Man Was Born*. Collins, 1972.

Schaller, George B., "Predators of the Serengeti" (Part 1: "The Social Carnivore"). *Natural History*, Feb. 1972.
The Deer and the Tiger. University of Chicago Press, 1967.

Van Lawick-Goodall, Hugo and Jane, *Innocent Killers*. Collins, 1970.

Fossils

Day, Michael, *Guide to Fossil Man*. Cassell, 1965.
Fossil Man. Hamlyn, 1969.

Isaac, Glynn, "The Diet of Early Man." *World Archaeology* (Vol. 2, No. 3), Feb. 1971.

Leakey, M. D., *Olduvai Gorge*, Vol. 3. Cambridge University Press, 1971.

Oakley, Kenneth, and Bernard Campbell, eds., *Catalogue of Fossil Hominids—Part I*. Trustees of the British Museum, 1967.

Tobias, Phillip V., *The Brain in Hominid Evolution*. Columbia University Press, 1972.

Primates

Altmann, Stuart and Jeanne, *Baboon Ecology*. University of Chicago Press, 1970.

Chance, Michael, and Clifford Jolly, *Social Groups of Monkeys, Apes, and Men*. Jonathan Cape, 1970.

DeVore, Irven, ed., *Primate Behavior*. Holt, Rinehart and Winston, 1965.

Jay, Phyllis C., *Primates*. Holt, Rinehart and Winston, 1968.

Jolly, Clifford, "The Seed-eaters." *Man* (Vol. 5, No. 1), March 1970.

Kummer, Hans, *Social Organization of Hamadryas Baboons*. University of Chicago Press, 1968.

Napier, John and P. H., *Handbook of Living Primates*. Academic Press, 1967.

Reynolds, Vernon, *Apes*. Cassell, 1964.

Schaller, George B., *Year of the Gorilla*. Penguin, 1967.

Schultz, Adolph H., *The Life of Primates*. Weidenfeld and Nicolson, 1969.

Simons, Elwyn L., *Primate Evolution*. Macmillan, 1972.

Van Lawick-Goodall, Jane, *In the Shadow of Man*. Collins, 1971.
My Friends the Wild Chimpanzees. National Geographic Society, 1967.

Index

*Numerals in italics indicate an illustration
of the subject mentioned.*

X

Typesetting by C. E. Dawkins (Typesetters) Ltd., London SE1
Printed and bound in Belgium by Brepols Fabrieken N.V.